PRIORITY SETTING *and* STRATEGIC SOURCING

in the Naval Research, Development, and Technology Infrastructure

Kenneth V. Saunders

Bruno W. Augenstein

Paul Bracken

Glenn Krumel

John Birkler

James Chiesa

Cullen M. Crain

R. Richard Heppe

Richard F. Hoglund

Brian Nichiporuk

Prepared for the
United States Navy
Office of the Secretary of Defense

RAND

National Defense Research Institute

This report suggests ways in which the Department of the Navy (DON) might realize more value from its increasingly constrained research, development, and technology (RD&T) dollars. The suggestions are presented in three parts. The first part develops and applies a framework for setting funding priorities in the Naval RD&T infrastructure. The second part discusses alternative RD&T procurement arrangements that are being used increasingly in the private sector and have been used in various parts of government; they are commonly called "smart buying," but we use the term "strategic sourcing." The third part is a speculative combination of the priority-setting and strategic-sourcing considerations of the first two parts; it suggests a way to help determine which parts of the Naval RD&T infrastructure are best suited for alternative procurement arrangements and a way to determine which facilities might be involved.

The research reported here was conducted to support DON policy formulation for the 1995 round of Base Realignment and Closure (BRAC) and to assist the Navy in meeting its longer-term need to make the best use of its resources. It was carried out in an intensive 9-week study effort that began early in January 1994. From March through June 1994, results were briefed to the Washington, D.C., parts of the Navy's research, development, and acquisition (RD&A) community and to selected officials in the Office of the Under Secretary of Defense (Acquisition and Technology).

The report is intended primarily for that community, but also for those outside it with an interest in Naval RD&T matters. We believe our priority-setting framework, although focused specifically on the Navy, is almost directly applicable to the other two military services. It would probably require some modification to be applicable to the Department of Defense (DoD) as a whole or to other parts of the government. Much of the material on strategic sourcing, however, applies as is, not only to DoD as a whole but to other parts of government as well.

The work was carried out in the Acquisition and Technology Policy Center of RAND's National Defense Research Institute, a federally funded research and development center sponsored by the Office of the Secretary of Defense, the Joint Staff, and the defense agencies. It was funded by the Deputy for Resources, Analysis and Policy (RA&P) in the Office of the Assistant Secretary of the Navy for Research, Development and Acquisition. It was part of a larger effort funded by RA&P, which additionally covered the engineering, production, and support parts of the Navy's infrastructure, and in which a number of other research organizations took part.

The cutoff date for information in the report is June 30, 1994.

CONTENTS

Part II: Devising New Sourcing Strategies

FIGURES

Many responsible elements of the Department of Defense (DoD), Congress, and the defense industry have expressed concern that the DoD, in downsizing and restructuring in recent years, is not cutting back on infrastructure, or non-combat elements, to a degree that is appropriate for the extent it is drawing down the combat force structure. Separately, proponents of the acquisition reform that is now under way question the need for DoD's maintaining as much infrastructure in-house as it has. Infrastructural elements are thus expected to get a more critical look in further rounds of budget-cutting. The Department of the Navy (DON) is undertaking such a critical examination—both to make appropriate preparations for the 1995 round of Base Realignment and Closure (BRAC) and to meet its need to make the best use of its resources for the longer term.

In its examination, DON has divided its infrastructure into two parts. One part comprises research, development, and technology (RD&T); the other part comprises engineering, production, and support (EP&S). The purpose of this study is to offer analysis that will aid DON in its review of its RD&T infrastructure. More specifically,

- we present a framework for setting priorities for funding different lines of RD&T, and we offer our best application of that framework (Part I)

- we discuss ways in which the Navy might increase the efficiency and effectiveness of its RD&T investments—mostly through flexible insourcing and outsourcing strategies, which are sometimes collectively referred to as "smart buying" or "strategic sourcing." We use the term "strategic sourcing" (Part II).

Our results on both topics have implications for deciding which facilities and other infrastructural elements should be retained and which should be eliminated. We generally leave it to DON to draw any conclusions it believes may be warranted about continuing or realigning specific facilities. The exception is in Part III—a speculative combination of the priority-setting framework and strategic-sourcing ideas—which names some specific institutions.

This report assumes that the policies and world view given in the DoD 1993 Bottom-Up Review (BUR) and two important, BUR-consistent Navy documents, . . . *From the Sea* and *Force 2001*, will not change significantly in the foreseeable future. Although this assumption does not necessarily drive the results, as an official DoD planning position following from the end of the Cold War, it was the "permeating atmosphere" in which the study was carried out. The report also tacitly assumes that any acquisition reform that would be needed to implement its suggestions will in fact occur.

PART I: SETTING PRIORITIES

The framework we developed to aid in establishing RD&T funding priorities comprises four steps: assembling a list of RD&T capabilities, defining criteria or dimensions along which the capabilities can be evaluated, ranking the capabilities according to the criteria, and translating the criteria rankings into support priorities. Because we regard the framework per se, rather than our best application of it, as the primary output of our priority-setting work, we emphasize it in this discussion. To arrive at final sets of RD&T funding priorities using this framework, DON must apply the framework itself.

Assembling a List of RD&T Capabilities

The first task is to assemble an appropriate (for priority-setting) list of RD&T lines, capabilities, or categories—we use the terms interchangeably—that are competing for DON funding. This task is not as straightforward as might be imagined. The list can be made arbitrarily short or long by lumping or splitting RD&T categories. A short list—one or two dozen entries—is perhaps more suitable for high-level priority-setting but may group capabilities so different that rankings would not be useful. Our final list contains 53 categories of RD&T (see Table 2.1 in the main text for an alphabetical listing). There is nothing magic about either the length of the list or its specific composition. If the Navy decides to apply the framework, it will probably end up with a list of different length and composition.

A question inextricably woven into this task is, What dimensions should be used in defining capabilities? Should RD&T capabilities be expressed in terms of capabilities that are operationally relevant in the system they support, the weapon systems to which they contribute, or the disciplines with which they are affiliated? We chose to include all three; thus, we have **Naval**

Oceanography, Ballistic Missiles, Ballistic-Missile Defense, Flexible Manufacturing, and **Fixed-Wing Aircraft** on the same list. Such inconsistent categorization actually facilitates setting priorities by clarifying what subject matter is included within each capability, and it reflects the untidy conflicts that actual priority-setters usually must face.

Defining Criteria

We selected two criteria for evaluating capabilities. The first is the relative value of future RD&T to the Navy. In our framework, value is judged primarily in relation to the contribution of the next dollar of funding to the ability of the Navy to achieve success in the next 20 to 30 years in its six joint mission areas: strike, littoral warfare, surveillance, strategic deterrence, space and electronic warfare, and strategic sealift. Various factors contribute to this judgment, including the potential for revolutionary capability improvements, speed of development, avoiding technological surprise, technological risk, the potential for multiple applications, and responsiveness to evolving threats.

By using *relative* value, we emphasize that there is a continual scale of value: There is no point or level above which a technology can be judged "critical"—all lines of RD&T are valuable. *Future* indicates that capabilities are not to be ranked on the basis of how valuable the line of RD&T has been to date, or on the value the RD&T that has already been accomplished or already exists will have in the future. Thus, for example, it is the value of future *RD&T* on, say, fixed-wing aircraft, that is assessed, not the value of existing and programmed aircraft per se.

Our second criterion is breadth of demand. We reason that, all other things being equal, one line of RD&T might be chosen for support over another if there are fewer demanders for the products of the former. For example, the Navy is the only demander for military-submarine RD&T, whereas helicopter RD&T is in demand not only in other parts of DoD but in the private sector as well. Thus, we established a scale of broadening demand (and thus supply), running from Navy-unique through DoD-only and government-only to cross-sectoral, i.e., generic. Because reconstituting a line of RD&T is difficult following shutdown, the Navy would be expected to think most carefully before terminating support for activities whose products only it will need. Again, here, we are interested in anticipated *future* demand.

Ranking the Capabilities

Our framework calls for establishing rankings by a Delphi method: building expert consensus among an appropriate group of experienced professionals, using rounds of "blind" individual ratings and rounds of open discussion of differences, then reevaluation. In applying this process ourselves, we found that to achieve consensus, we occasionally needed to be able to redefine categories and change their number as we went along. So it is important that the list not be "set in concrete" before ranking begins. For the same reason, we also found that we sometimes needed to be able to establish borderline values on the criterion scales. So it is also important to permit flexibility in this respect during the process.

A key to successful application of the framework by the Navy will be to carry out final list-making and ranking at a high enough level that participants' inputs and votes are driven primarily by interests no narrower than the Navy as a whole. Otherwise, given intra-Navy competition, categories centered around favored projects would result. Selective use of outside consultants and retired high-level military and civilian Navy personnel could facilitate input of expert information.

Our best application of the first three steps of the framework is shown in Figure S.1, which presents it on an orthogonal plot of the two criteria. The figure shows four main *levels* of relative value, running down from A, the highest, to D, as well as three borderline levels, AB, BC, and CD. It also shows four *classes* of breadth of demand and indicates one borderline class, between classes 1 and 2, by placing one of the circles on the line between those classes. No meaning is attached to the location of capability circles *within* an individual cell (the intersection of a class and a level); no capability in an individual cell is intended to be ranked higher or lower than any other capability in that cell.

We emphasize that the indicated list and these rankings are only *our* best application of the framework. Only the Navy itself can make final applications. At the most, our application will be corroborated by DON's own; at the least, it is intended to provoke discussion and thereby give some sense of the trade-offs and compromises that must be made in this kind of exercise.

Three important trends are apparent in the figure. First, RD&T categories judged of greatest value (level A) tend to be associated, directly or indirectly, with what is commonly called "the information revolution," i.e., computers and communications. Second, RD&T associated with the three main types of classical combatant platforms—submarines, surface ships, and fixed-wing aircraft—fall into relative-value class BC, mainly because the current platforms of these types are already so good. These two trends reflect our judgment that, generally speaking, more value will accrue in the next 20–30 years from things that will be installed in, hung on, or communicate with these types of platforms than will come from advances in the platforms per se. Third, RD&T categories associated with basic research tend to be located toward the lower right, a

RAND*MR588-S.1*

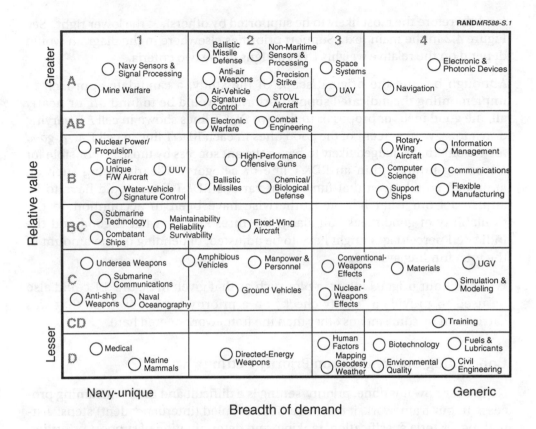

Figure S.1—Rankings of Naval RD&T Capabilities

situation that results from our choice of a 20–30-year time frame (a short one for basic research) for assessing value and our focus being strictly limited to Naval considerations. That situation does not reflect our strong conviction about the highest importance and absolutely certain longer-term payoff of basic research for the broadest national security and national interests.

Establishing Support Priorities

The plot in Figure S.1 is intended to aid DON in setting funding (or support) priorities for the various lines of RD&T. Although we are even less well-positioned to suggest support priorities than to rank capabilities, it appears obvious that capabilities with the highest relative value and the narrowest breadth of demand (or market demand) would be of highest funding, or support, priority. These capabilities lie at the upper left of the diagram. The lowest-priority areas would be the least-valuable ones with the widest demand

(and therefore the most likely to be supported by others), at the lower right. See Figure 4.2 in the main text. Support priorities elsewhere in the diagram would depend on the relative weights DON assigns to the two criteria.

Although beyond the original scope of the study, a candidate approach to implementing the indicated support priorities would be to fund *all,* or nearly all, the good ideas or proposals in the two RD&T lines shown in cell A1, varying down to, say, 10 percent of the good ones in each RD&T line in cell D4. A *good idea* is one that is judged likely to have relevant success by those responsible for allocating funds within an RD&T line or category. Clearly, using such an approach could mean that funding in various RD&T lines would have to be adjusted from what it would otherwise have been to accommodate the availability of good ideas. Similarly, the threshold for what is good and the indicated percentages might have to be adjusted, depending on the amount of the total funding available.

While not our original intent, a plot such as that given in Figure S.1 could also be used to provide a "sanity check" on a priority-setting exercise that was carried out by some means other than the framework offered here.

Concluding Observations on Priority-Setting

No matter how it is done, priority-setting is a difficult and time-consuming process. In our framework, it comprises four coupled (interdependent) steps: list-making, criteria specification, ranking, and determination of support priorities. Ostensibly sequential, these steps are, in fact, highly iterative. The value criterion is relative, thereby avoiding the more commonly used critical/noncritical dichotomy. The other criterion, breadth of demand for products of the Navy's RD&T infrastructure, captures the potential for overlap in capability, thereby opening the door to the question, Might the products of various lines be available elsewhere?

The ranking, and the other steps also (because of the iterative nature of the process overall), must be done at high enough levels of the Navy that the parties involved have the interests of the Navy as a whole as their top priority. It might be appropriate, for input of information only, to include outside experts and retired Naval personnel at intermediate stages of ranking.

We emphasize that basic research does not come out as a top priority by our criteria, because the time period for payoff—20–30 years—is so short and because our focus is limited to the Navy. Nevertheless, our strong conviction is that basic research is of highest national importance and will have absolutely certain longer-term payoff **some**time, **some**where, **some**how.

We also emphasize that we regard the framework itself as the primary output of the priority-setting part of the report. Our application of the framework should be viewed as our best cut, intended as much to provoke debate as to achieve concurrence. We do not have sufficient Naval expertise to confidently advise DON on what its RD&T priorities should actually be. We encourage DON to use the framework offered here as a means of structuring priority-setting at high levels within the department.

PART II: DEVISING NEW SOURCING STRATEGIES

We now turn our attention from deciding what to buy to deciding how and where to buy it, i.e., how and where to "source" it. We draw lessons from both the private and public sectors on what we refer to as *strategic sourcing*—the process of taking a longer-term and bigger-picture view of sourcing. The process involves deciding what is most important to the sourcing organization, then addressing in that context the associated decisions about how and where to source.

In the private sector, the boundary lines between organizations are being rethought. In the past, these lines were sharp; now, they are becoming blurred. We believe it is important for the Navy to understand this change.

Firms are inducing competition among their internal elements while establishing long-term partnerships with external firms, elevating suppliers to a higher role and status as they attempt to shape the supplier market. They are also taking steps to manage political risks—risks that follow from the actions of one's own government. One such evolving action is the move toward shifting health-care costs from government to employers. Employers, in turn, are shifting parts of their workforces to temporary and contingent kinds, in part to reduce the liability associated with these costs. DON cannot, of course, go this far in considering more flexible, strategic sourcing, nor would it necessarily want to; but any serious thought along those lines could have the same benefit it has often had in the private sector: that of revitalizing the organization—of returning its focus to its central, or core, mission.

In the public sector, some specialized DoD organizations or programs have realized significant innovations in managing certain sourcing activities. The examples we give demonstrate that, even though DoD differs from the private sector in history, culture, missions, and managerial structures, flexibility in sourcing approaches can be achieved—and has been—and can be effective.

The public-sector examples support a major theme that is worth stating: Technology alone and in itself is generally not the key to the desired answer;

rather, having the skill and imagination needed to apply that technology to an *integrated* system for a *specific mission* gives rise to the real accomplishment.

Private-Sector Case Studies

Eastman Kodak and Cummins Engine are two illustrative examples of the strategic-sourcing process. In 1989, because of competition and advancing technology, Kodak faced the decision of whether to modernize and consolidate its varied information-processing technology (IT) system components. The firm decided, instead, to concentrate its capital and efforts on such core capabilities as digital imaging and specialty chemicals and to outsource most of its IT services. Outsourcing required both transferring personnel to the vendor and involving the vendor intimately in internal Kodak operations—radical changes in Kodak's corporate culture, part of which was that a job there was a job for life. As a major example of outsourcing complex corporate operations, the Kodak example makes the point that how outsourcing is done—the planning and managing of the transition—is just as important as the decision to do it in the first place.

As with Kodak, Cummins was forced by competition to become more efficient. And as with Kodak, instead of trying to optimize across the board, Cummins began outsourcing noncore capabilities. Senior decisionmakers faced charges, stemming from concerns felt by various quarters within the organization, that they were "hollowing out" the firm and that it would be difficult to decide what was so important it could not be outsourced. These concerns revealed a lack of strategic framework, a failure to realize what activities actually added value to the firm's principal products. Decisionmaking had too often been an exercise in compromising among organizational constituencies, some of them peripheral to the central mission. More outsourcing on DON's part could well lead to similar concerns, amelioration of which can perhaps be effected by drawing on these lessons-learned from private-sector experience.

Public-Sector Case Studies

Whereas the private sector has had more experience in flexible sourcing than the public sector, some instances in the latter have particular relevance to DON. The Advanced Research Projects Agency (ARPA) represents an organization that has funded a great deal of high-payoff R&D with virtually no in-house R&D infrastructure of its own. (We recognize that ARPA sometimes makes good use of the services' in-house infrastructure.) DON would not, of course, want to do the same, but ARPA offers some useful lessons. Its minimal bureaucracy and extensive autonomy for managers help in attracting very good people from universities, industry, and elsewhere in government. Its funded activities are characterized by a balance between exploratory R&D and system development. But regardless of where an activity is located in the R&D spectrum, a premium is placed on the need for applicability in some operating system.

A second public-sector example is that of compartmented programs (also referred to as "black programs"). Such programs are characterized by intimate buyer-contractor relations based on trust, by a focus on system development, and by a small management staff with little service-laboratory involvement. (We recognize that there is a downside to black programs, e.g., sometimes insufficient oversight and peer review.) DON might benefit by going part of the way along each of these dimensions in its own RD&T acquisition.

Although the airline industry is not in the public sector, its sourcing strategies are similar to those of compartmented programs, and airlines and DoD act as buyers from the same set of aircraft-industry suppliers. As with compartmented programs, the airline industry has no in-house laboratories and supports hardly any R&D. It expects the suppliers to provide a virtually complete, ready-to-operate product. The buyer does not micro-manage the suppliers' activities and can therefore maintain a smaller procurement organization. Nevertheless, mutual interests result in a great deal of interaction between buyers and suppliers, and the two function almost as partners.

The traditional categorization of defense R&D (6.1, 6.2, etc.) does not serve well to emphasize the value added through integration and mission application. Because of that categorization, and because defense RD&T actors are usually assigned to discrete activities within those categories, there is little freedom or opportunity for life-cycle involvement in providing value-added and affordability to a user. "Smart buyers" would be specifically empowered to maximize the advantages inherent in such attributes.

Efficiency Gains from Use of Civil-Sector Production

The preceding examples suggest ways in which DON might consider sourcing decisions more strategically, with the possible result of transferring more of its RD&T capability outside the organization. But in addition to finding more strategically sound, efficient ways to manage RD&T, a more flexible approach could also sometimes entail buying the products of commercial RD&T "off the shelf." In such instances, there would be little or no investment or involvement in the RD&T itself.

For example, most funds invested recently in microelectronics R&D have come from the commercial sector. Technologically advanced commercial micro-processors and other products have equaled or led military products in speed of introduction, performance, and performance per dollar. Were it not for specialized military packaging and testing requirements, DoD could, in most instances, have bought such products instead of developing its own, and would have been just as well off. The utility of such requirements should be scrutinized in as much as they drive up DoD's costs and act as a drag on the whole industry, because separate production lines must be sustained.

This situation was brought into sharp focus with DoD's 1980s program to de-velop very-high-speed integrated circuits. DoD hoped to push the state of the art well ahead of the commercial sector, and it set barriers to commercialization in an effort to keep the technology out of Soviet hands. But, motivated by com-mercial demand, firms that either did not wish to accept such constraints or that were eliminated from the DoD competition went ahead on their own and produced products with similar performance over a similar time frame.

The advantages to be gained from dual-use production lines have been amply demonstrated by the Japanese. For the FS-X fighter aircraft, which they are de-veloping collaboratively with the United States, the Japanese have designed on their own an advanced phased-array radar that makes extensive use of gallium arsenide (GaAs) technology. The Japanese have been developing that technol-ogy for commercial purposes for over 20 years, and they are the major supplier of the world's demand for that technology.

The radar's performance characteristics are not superior to those of the radar the United States is now developing for its own F-22, which also uses GaAs technology. In fact, they will, apparently deliberately, be less so. But that is not the point. The Japanese radar will be much less expensive to produce than the U.S. radar—because it takes advantage of Japanese integration of military R&D and production with R&D and production aimed at the larger commercial market. The Japanese have given up some performance to get something that is much more affordable. The U.S. approach of separating military and civilian development and production results in the surrender of dual-use producibility advantages.

Even in the field of intelligence-gathering, the economic and technological value of exploiting commercial resources will grow rapidly. Virtually every country of interest has vital information available on-line, including maps and mapping imagery, technical and prototype engineering data, information on economic and development trends, and data on military structure and systems. Clearly, commercial databases will be an increasingly powerful tool for military and national intelligence, and for threat assessments.

Implementation Issues

The success of any transition to more strategic and flexible sourcing would depend on how the transition was implemented. Both leadership and management are important, as is recognizing the distinction between them. *Leadership* deals with change; *management* deals with the complexity of implementing changes that leadership points to. Both leadership and management are needed to change an organization. Leadership without management is empty and abstract. Management without leadership is unexciting, uninspiring, and unfocused. It takes leadership to see sourcing as strategic; for example, to go beyond consulting only the management dimensions of an organization—which will probably advise against change.

DON might consider several private-sector implementation aids. Special transition teams might be set up with decisionmaking power (and responsibility), but they would last only as long as the transition, to minimize long-lasting institutional resentments. In areas where infrastructure is given up, teams of experts, called *architectural knowledge teams*, could be retained in-house to help ensure "smart" buying. *Fluid contracting*, which emphasizes long-term joint trust rather than litigation threats, could also be employed to enhance short-term performance. Finally, various steps might be taken to enhance communication of information and concerns within the organization and to educate mid-level managers in new responsibilities.

Concluding Observations on Strategic Sourcing

The private sector in the United States has greatly increased its use of outsourcing. The strategic-sourcing part of this report argues that the underlying rationale behind this trend is much broader than short-term cost savings. There are many reasons for outsourcing, ranging from the constructive dismantling of existing organizations and corporate cultures to reducing or eliminating political risks.

It is important for DON to understand that the reasons behind this trend are many and varied. If the strategic-sourcing part of this report does nothing else, it should stretch DON's thinking about rationales for outsourcing. In this particular field of public policy analysis, it is striking that the most important reasons for outsourcing in the private sector appear to have been debated little in the Navy.

Another conclusion that comes out of this work will be harder to implement than expanding DON's thinking on *reasons* for outsourcing. Outsourcing comes from an understanding of *why* an organization sources the way that it does. DON would have to ask itself what is most important to it and what is not.

Things that are not core are the more promising candidates for outsourcing. Things that *are* core to it can be done either inside or outside, depending on many variables. In a field such as R&D, many examples suggest that it is possible to keep so much work in-house that contact with new ideas and technologies is eroded.

Strategic sourcing can be a therapeutic management exercise with unexpected positive results. Its application to DON will be imperfect, true, but in attempting to apply it, department decisionmakers will be focusing on truly key areas of what the Navy does and how it does it.

PART III: SUITABILITY OF RD&T LINES FOR OUTSOURCING AND POSSIBLE IMPLICATIONS OF THE FRAMEWORK FOR THE NAVAL RD&T INFRASTRUCTURE

We speculate that the priority-setting framework and strategic-sourcing concepts can be combined and used to draw inferences about which lines of RD&T are better suited for outsourcing. In principle, even high-relative-value, Navy-unique lines of RD&T could be outsourced. However, in practice, a better case can be made for retaining RD&T capability in-house for those lines for which it is anticipated that future RD&T will have high relative value than can be made for those lines for which the value is judged to be relatively less. At the same time, outsourcing options would be more varied when the anticipated future demand (and therefore supply) is broader.

Figure S.2 is the same as Figure S.1, except that the circles are not identified—so that we can concentrate on the ideas—and the shading is different. This reinterpretation of Figure S.1 (which is intended as an aid in setting funding priorities) might be of use in deciding which lines of RD&T can be more flexibly sourced. Thus, regions of increasing suitability for flexible sourcing might resemble those represented in Figure S.2 by the increasingly lighter shading.

Two provisos should be attached, however: First, it is much clearer that RD&T categories toward the upper left need to be supported *somehow* than that they need to be supported in-house. Second, when reliance on commercial-sector products is specifically sought, it follows from the definition of classes of breadth of demand that the only suitable candidates are those on the right side of the diagram.

A finished chart like that in Figure S.2 might also provide a basis for guiding decisions about where flexible sourcing could substitute either partially or completely for elements of the in-house RD&T infrastructure. The more those lines of RD&T positioned toward the lower right of the diagram were already

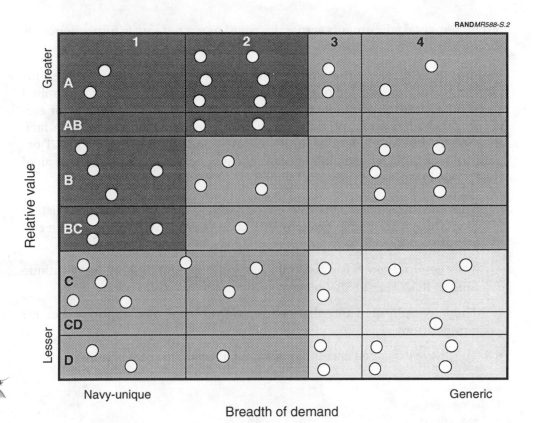

NOTE: The lighter the shading is, the greater the suitability for flexible sourcing.

Figure S.2—Inferring Suitability for Flexible Sourcing from Relative Value and Breadth of Demand

represented by institutionally or physically coherent elements of the infra-structure, the clearer the guidance.

Thus, our framework might be extended to support decisionmaking on the continuation of research facilities as integral parts of the Naval RD&T infrastructure. This extension could identify those facilities that, by pursuing the highest-relative-value and narrowest-demand technologies, warrant the highest support priority. It could also highlight those that warrant a lower priority.

Another approach would be to retain those elements most oriented to system development, i.e., put research, exploratory development, and most or all of advanced development on a more flexible sourcing basis. Certain key capabilities would be retained, but most would probably be outsourced. Exceptions would have to be specifically defended. This approach might seem Draconian, but incremental adjustments across the board might not satisfy future demands for infrastructure downsizing.

And it is quite plausible that there would be defensible exceptions. For example, the Naval Research Laboratory and The Johns Hopkins University Applied Physics Laboratory are highly productive, world-class institutions, major elements of which it would probably be in DON's interests to preserve. In fact, they might serve as models for future in-house and nongovernment RD&T organizations. Such retained infrastructural elements should have some or all of the following characteristics:

- They have a retained corporate memory and a sense of institutional history (not always present in government organizations with high turnover in senior positions).

- They conduct core basic research, with participation in select Navy-unique lines of RD&T (e.g., SSBN security).

- They are subjected to enhanced program review and opportunities for competition.

- They have expanded access of personnel to academic career paths.

ACKNOWLEDGMENTS

The authors wish to acknowledge the outstanding support, throughout the course of the study, of the project monitor, Ms. Terry Whalen, then of the Office of the Assistant Secretary of the Navy for Research, Development and Acquisition, under the Deputy for Resources, Analysis and Policy (ASN/RD&A/RA&P). They also wish to acknowledge the interest and valuable guidance of RADM Dave Oliver, then Deputy for RA&P in the OASN/RD&A, who initiated the study.

Additionally, they thank Philip Selwyn, a former director of the (then) Office of Naval Technology, for bringing the Tri-Service S&T Reliance work to their attention; Matt Jaskiewicz, then Executive Secretary of the Joint Directors of Laboratories, for his valuable discussions and splendid cooperation in getting Reliance documentation for us; Paul Wessel, Technical Director of the Naval Command, Control and Ocean Surveillance Center, for valuable comments on some of our early work; and the many Navy and other government officials for valuable feedback on our final results during the briefing cycle. Four of the latter offered particularly constructive suggestions: RADM Walt Cantrell, Commander, Space and Naval Warfare Systems Command; Timothy Coffee, Director of Research, Naval Research Laboratory; RADM Oliver; and RADM Don Pilling, Director, Programming Division, Office of the Deputy Chief of Naval Operations (Resources, Warfare Requirements, and Assessment).

We are also grateful for the valuable comments and suggestions from RAND colleagues Gene Gritton, Dick Hundley, Michael Kennedy, Glenn Kent, John Schrader, Giles Smith, and Willis Ware; the insightful and useful work of the formal reviewers, Jim Dewar and Elwyn Harris; and the yeoman contributions, each in her own way, of the editor, Marian Branch, and the secretaries, Sabrina Cowans and June Kobashigawa.

Of course, the authors are responsible for the final content of the document.

ACRONYMS AND ABBREVIATIONS

ACAT	Acquisition Category
ADCAP	Advanced Capability
ADVCAP	Advanced Capability
AOE	Fast Combat-Support Ship
ARPA	Advanced Research Projects Agency
ASUW	Anti-Surface Warfare
ASW	Anti-Submarine Warfare
BRAC	Base Realignment and Closure [Commission]
BUR	Bottom-Up Review
C3CM	Command, Control, and Communications Countermeasures
CFD	Computational Fluid Dynamics
CNO	Chief of Naval Operations
CNR	Chief of Naval Research
DNA	Defense Nuclear Agency
DoD	Department of Defense
DON	Department of the Navy
DSB	Defense Science Board
DSPs	Digital signal processors
EHF	Extremely high frequency
EMP	Electromagnetic pulse
EO	Electro-optics
EP&S	Engineering, Production and Support
EW	Electronic warfare
FSU	Former Soviet Union
GaAs	Gallium arsenide
GPS	Global Positioning System
HR	Human Resources

IC	Integrated circuit
ID	Identification
IT	Information technology
IR	Infrared
JAST	Joint Advanced Strike Technology [Program]
JCS	Joint Chiefs of Staff
JDL	Joint Directors of Laboratories
LAMPS	Light Airborne Multipurpose System
LSD	Landing Ship Dock
MC&G	Mapping, Charting, and Geodesy
MM	Millimeter
MRC	Major regional conflict
NASA	National Aeronautics and Space Administration
NAVAIR	Naval Air Systems Command
NAVSEA	Naval Sea Systems Command
NAVSO	Naval Supply Office
NAVSUP	Naval Supply Systems Command
NAWC	Naval Air Warfare Center
NCCOSC	Naval Command, Control and Ocean Surveillance Center
NLCCG	Navy Laboratory/Center Coordinating Group
NSWC	Naval Surface Warfare Center
NRL	Naval Research Laboratory
NUWC	Naval Undersea Warfare Center
OCNR	Office of the Chief of Naval Research
OSD	Office of the Secretary of Defense
PEN	Program Element Number
PIP	Product Improvement Program
RA&P	Resources, Analysis and Policy
R&D	Research and Development
RD&A	Research, Development and Acquisition
RD&T	Research, Development and Technology
RDT&E	Research, Development, Test and Evaluation
RF	Radio frequency
RISC	Reduced-instruction-set computer
RPG	Receiver Processor Group
ROMs	Programmable read-only memories
SLBM	Submarine-Launched Ballistic Missile

SPAWAR	Space and Naval Warfare Systems Command
SQUID	Superconducting Quantum-Interference Device
SRAMs	Static random-access memories
SSBN	Nuclear-powered ballistic-missile-carrying submarine
SSN	Nuclear-powered attack submarine
SSPO	Strategic Systems Project Office
S&T	Science and Technology
STOVL	Short Takeoff and Vertical Landing
SURTASS	Surveillance Towed-Array Sonar System
UHF	Ultra-high frequency
VHF	Very-high frequency
VHSIC	Very-high-speed integrated circuit
VLSI	Very-large-scale integration

Chapter One

INTRODUCTION

BACKGROUND

Since the mid-to-late 1980s, the Department of Defense (DoD) has been downsizing and restructuring. The process began accelerating with the end of the Cold War, and, unless conditions change, is expected to continue into the next century. It is affecting both the force structure, or combat forces, and the associated infrastructure, which are often referred to as DoD's "tooth" and "tail," respectively. There is general agreement among many responsible DoD and other government officials, and in the defense industry, that downsizing of the infrastructure *is* lagging that of the force structure, with the attendant danger of, eventually, "the tail wagging the dog." Thus, the DoD is looking at its infrastructure with an eye toward significant further downsizing and restructuring. Possible infrastructure reductions have been the subject of considerable press attention[1] and at least one dedicated conference.[2]

The Department of the Navy (DON) expressed similar concerns about reductions in its infrastructure in a 1993 publicly released briefing.[3] The DON is worried that, if financial savings from infrastructure reductions are not realized while total department funding *independently* continues falling, funding for the operational forces will necessarily decrease even further than planned to make up the difference. That is to say, the Navy's tooth-to-tail ratio will decrease still further.

---

[1]See Thomas E. Ricks and Michael K. Frisby, "Clinton Nominates Perry for Defense Secretary, but Still Faces Doubts About Strength of His Team," *Wall Street Journal*, January 25, 1994.

[2]See Samuel Kleinman and Carla Tighe, eds., *Shrinking the Defense Infrastructure*, Alexandria, Va.: Center for Naval Analyses, Conference Report, 1993.

[3]See U.S. Navy, Chief of Naval Operations, *Restructuring Naval Forces for New Challenges: The FY95–99 Navy Program Review*, and associated briefing charts, Washington, D.C.: PR-95, November 1993.

The current concerns and the planning efforts being undertaken to address them arise partly from previous experiences with downsizings, i.e., those following World War II and the Korean and Vietnam conflicts. In the atmosphere of war-weariness that naturally prevailed at those junctures, sufficient attention was not always paid to rationally downsize operational forces and their supporting infrastructure while preserving the capability to react to changes in the world situation. Nobody wants to repeat such mistakes.

Planning for rational downsizing is further motivated by the recognition that DoD acquisition reform is starting. That reform is warranted by several factors, among which are the private sector's technological lead in areas associated with the information revolution—computers and communications—and the realization that, in many cases, DoD might be able to both save money and get better products sooner by buying commercially. That realization has led to a questioning of the need for significant amounts of the DoD infrastructure, thus reinforcing other motivations for infrastructure downsizing.

Complicating the picture is the natural tension in Congress between demanding DoD downsizing and attempting to preserve public-sector jobs.[4] Recognizing this tension, Congress in 1988 created the Base Realignment and Closure (BRAC) Commission to attempt to minimize the effects of politics in infrastructure reduction. Given the BRAC Commission's power, elements of DoD are coming to embrace the potential of careful planning for transforming what could be termed the BRAC "meat ax" into something more like a skillfully wielded knife. One key to realizing such a transformation is setting priorities among infrastructure activities, with the goal of using the results to inform recommendations to the BRAC. That way, DoD and the individual services could attempt to maximize the value to themselves of the parts of the infrastructure that are retained.

The Office of the Assistant Secretary of the Navy for Research, Development and Acquisition (OASN/RD&A) is responsible for many of the Navy's preparations for the 1995 BRAC round and for longer-term planning. These activities include examinations of the total Naval infrastructure, divided into two parts: (1) research, development, and technology (RD&T), and (2) engineering, production, and support (EP&S).[5]

---

[4]See John F. Harris, "Defense Cuts: Not in My District," *Washington Post*, June 9, 1994.

[5]This appears to be a functional division rather than one made along budget or organizational boundaries.

PURPOSE AND SCOPE

The purpose of this report is to provide the ASN/RDA with analysis of the RD&T portion in support of both the Navy's BRAC 95 policy-formulation process and its longer-term planning needs. We do not deal with the EP&S parts of the Navy's infrastructure.[6] Moreover, we do not generally deal with specific facilities or with budget issues and information. Rather, we make suggestions for ways the Navy might realize maximum value from its available RD&T dollars through two kinds of activities: setting RD&T support priorities and adopting alternative RD&T procurement, or *sourcing*, approaches. More specifically, we

- present a framework for setting support priorities among the myriad of Naval RD&T capabilities and offer our "best cut" at applying the framework

- discuss ways in which the Navy might increase the efficiency and effectiveness of its RD&T investments—mostly through more flexible insourcing and outsourcing strategies, which are sometimes collectively referred to as "smart buying" or "strategic sourcing."[7] We use the term "strategic sourcing."

Of course, it may be possible to draw inferences from the study outputs about the continuation, downsizing, or elimination of specific elements of the infrastructure. DON is in the best position to draw such inferences; we generally do not attempt to do so here. The exception involves our speculative combining of the priority-setting framework and strategic-sourcing ideas in Chapter Twelve, which names some specific institutions.

In performing the analysis, we assumed at the outset that the DoD world view and policies given in the 1993 "Bottom-Up Review" (BUR) briefing[8] will not change significantly in the foreseeable future.[9] Our assumptions are that

---

[6]Other independent research organizations are addressing EP&S. However, we are not the only ones performing RD&T analyses. The Navy's four warfare centers, for example, are conducting investigations of their own infrastructures, both in preparation for the 1995 BRAC round and for longer-term planning appropriate to their missions. In contrast, this report addresses the aggregate DON level.

[7]There are other ways to attempt to increase efficiency and effectiveness besides changing sourcing policy. Sourcing was chosen as our focus mainly because, in our view, the Navy is not as cognizant of sourcing as it is of other ways. Additionally, it offers great potential and is timely, given the current relevance of the topic in both academia and the private sector. We believe it is particularly appropriate in the current downsizing environment.

[8]See Les Aspin and General Colin Powell, "Bottom-Up Review," Washington, D.C.: U.S. Department of Defense, briefing charts, September 1, 1993.

[9]This assumption provided the "permeating atmosphere" in which the whole study was carried out. We invoked it because it is "official," not because we, or RAND as a whole, necessarily agree with it. The environment resulting from the end of the Cold War could more reasonably be considered the real permeating atmosphere; the BUR is merely one of many possible alternative DoD planning positions that could have followed. It is not clear to us whether all or even most of the study results

- the former Soviet Union (FSU) will not be reconstituted into something resembling what it was during the Cold War, but the possibility must be hedged against

- the roles and missions of the services will remain essentially as delineated in former Joint Chiefs of Staff (JCS) Chairman Colin Powell's 1993 review[10]

- the BUR's force levels will apply for the foreseeable future, as will its goal of being able to win two nearly simultaneous major regional conflicts (MRCs).

Not only did we accept these assumptions as stated, we did not attempt to analyze the sensitivity of our results to possible variations from the BUR's conclusions.[11]

We also accepted the assumptions and policies promulgated in two recent, important Navy documents, . . . *From the Sea*[12] and *Force 2001*.[13] The former is the Navy's first and highest-level, post–Cold War policy statement. The latter, issued about a year later, affirms the earlier statement and explains the associated Navy programs. Both documents appear to be consistent with the BUR.

These scope limitations mean that we do not treat several important, far-ranging issues that are relevant to Naval RD&T investment planning:

- Whether there will soon be significant changes in the Navy's roles and missions vis-à-vis those of the other services,[14] and even if there are not,

---

are driven by the assumption. There is still ambiguity and uncertainty about what exactly drove what. Ostensibly, the assumption has little impact or constraining effect on the work except in Chapter Four. But, as is shown below, because of the necessarily iterative nature of the process of applying the framework, it therefore could affect Chapters Two and Three as well. Similarly, although it is even less obvious that the assumption drives the sourcing part of the study, the assumption again was the "permeating atmosphere" in which the work was carried out, and thus we deem it appropriate to state it here. We briefly return to the "permeating-atmosphere question" in footnote 5 of Chapter Four.

[10]Colin L. Powell, Chairman, Joint Chiefs of Staff, *Chairman of the Joint Chiefs of Staff Report on the Roles, Missions, and Functions of the Armed Forces of the United States*, Washington, D.C.: U.S. Department of Defense, CM-1584-93, February 10, 1993.

[11]Having said this, we note that it is easy to see that there could be great sensitivities. Anticipating some of the results in Chapter Four, for example, if, instead of merely hedging against the possible resurgence of the FSU, we assumed it would be reconstituted and the Cold War would resume, the relative value of **Submarine Technology** would almost certainly be greater than it is there.

[12]See U.S. Navy, Chief of Naval Operations, . . . *From the Sea: Preparing the Naval Service for the 21st Century*, Washington, D.C., September 1992.

[13]See U.S. Navy, Chief of Naval Operations, *Force 2001: A Program Guide to the U.S. Navy*, Washington, D.C., July 1993.

[14]There have been numerous calls to revisit former JCS Chairman Colin Powell's 1993 review of service roles and missions. See, e.g., Steven Watkins, "Bombs Away! McPeak Eyes the Navy's Air Mission," *Air Force Times*, March 2, 1994. Congress responded to these calls, in the language of the National Defense Authorization Act for Fiscal Year 1994, by authorizing the creation of the

- Whether the BUR requirement of winning two nearly simultaneous MRCs is appropriate,[15] and if it is,

- Whether sufficient funds will be budgeted to buy the BUR forces,[16] and if they are,

- Whether BUR force levels will be sufficient to meet the BUR requirement of winning two nearly simultaneous MRCs.

APPROACH

The report is assembled in three parts. The approach for each part is described below.

Part I—Setting Priorities

If Naval RD&T funding falls substantially in the coming years, as now appears possible, the DON will probably be continually revisiting the support priorities it attaches to various lines of research. Doing so will not be easy, because all lines of research are important in one way or another—or they would not currently be supported—and all will have proponents. But the alternative to revisiting priorities is imposing across-the-board spending cuts, which could mean some lines of research will not advance enough to bear fruit. In the first part of the report, we propose a framework for setting Naval RD&T priorities in a downsizing and less-threatening environment,[17] and take our "best cut" at applying it.

We emphasize that we regard the framework itself as the primary output of the work documented in the first part. Our application of the framework should be viewed only as our best cut, presented for concreteness and as much to provoke debate as to achieve concurrence. We do not have sufficient Naval expertise to confidently advise DON on what its RD&T priorities should be.

---

independent Commission on Roles and Missions of the Armed Forces. As of the cutoff date for information for this report (June 30, 1994), things were actually in motion for what appears will be a FY 95 effort by the indicated commission.

[15]See, e.g., Eric Schmitt, "Cost-Minded Lawmakers Are Challenging a 2-War Doctrine," *New York Times*, March 10, 1994, p. 2.

[16]See, e.g., "Senate Panel Gives Clinton B-2 Funds He Didn't Seek," *Congressional Quarterly*, June 11, 1994.

[17]Less threatening in that the Cold War is over and, thus, short-notice threats (a few hours to a few days or weeks) to the very existence of the United States are significantly less. This is not to suggest that, in fact, the world is less threatening in other ways.

Because the methodology is the main product of this part of the work, we have organized it according to the four main steps we took in its development and application:

- Assembling a candidate list of Naval RD&T capabilities (discussed in Chapter Two)

- Selecting criteria for ranking the list entries (Chapter Three)

- Ranking the entries according to the criteria (Chapter Four)

- Developing a framework for setting support priorities that uses the ranked lists and criteria (also Chapter Four).

Despite the orderly progression of steps indicated by this sequence, the reader should bear in mind that the actual process was iterative. Chapter Five recapitulates and gives concluding observations on the priority-setting part of the report.

Part II—Devising New Sourcing Strategies

If RD&T funding is to be cut, and DON needs to curtail support for some lines of effort, it will obviously be to the department's advantage to maximize the benefit from the remaining dollars. More efficient and effective RD&T procurement could enable DON to meet infrastructure budget targets with less loss of potential future capabilities. Greater efficiency and effectiveness might come in the form of improved processes or a leaner infrastructure.

Within the private sector, determining procurement sources has become more of a strategic endeavor, requiring the firm to focus on its strengths, consider the spectrum of possible internal and external sources, and establish a more cooperative relationship with outside suppliers. We review the general characteristics of strategic sourcing in some detail (Chapter Six) and discuss case studies, both in the private sector (Chapter Seven) and in the public sector (Chapter Eight). We consider a full range of sourcing options, from alternative ways of procuring RD&T services to purchasing finished, commercial, off-the-shelf products (in Chapter Nine). In Chapter Ten, we discuss institutional and other considerations in implementing alternative sourcing strategies. In Chapter Eleven, we conclude Part II with some overall observations on strategic sourcing.

Part III—Combining Parts I and II and Drawing Inferences

In Chapter Twelve, we return to the priority-setting framework and make a speculative attempt to combine it with strategic-sourcing concepts and ideas, thereby inferring suggestions for DON sourcing strategies. Specifically, we indicate how inferences might be drawn about which RD&T activities are the better candidates for outsourcing and sketch a methodology for determining which infrastructural elements should be retained or realigned.

PART I

SETTING PRIORITIES

Chapter Two

ASSEMBLING A LIST OF NAVAL RD&T CAPABILITIES

In this chapter, we develop our list of Naval RD&T capabilities, categories, or lines—terms we use interchangeably. We first explain what we mean by "RD&T capability" and what constitutes our understanding of the Navy's in-house RD&T organizational structure. We then describe our candidate sources of information about Naval RD&T capabilities and discuss our final list. The process of arriving at that list was not trivial. Because it may provide useful lessons for anyone applying our methodology, we discuss the process as we move through the chapter.

NAVAL RD&T AND THE ORGANIZATIONS INVOLVED

Whereas specific budgetary considerations are outside our scope of analysis, some discussion of the current budgetary structure is helpful in explaining what we include as Naval RD&T:

- Activities carried out under what the Navy calls Science and Technology (S&T),[1] i.e., activities that are funded under Budget Categories 6.1, 6.2, and 6.3A (Research, Exploratory Development, and Advanced Technology Development)

- Some activities funded under Budget Category 6.3B (Demonstration and Validation[2])

- Some activities under the other ten DoD Programs.[3]

Naturally, we include management of the RD&T activities within our scope. Although acquisition milestone 1, a natural RD&T/EP&S demarcation, occurs

---

[1]U.S. Department of the Navy, *RD&A Management Guide*, 12th ed., Washington, D.C.: U.S. Government Printing Office (GPO), NAVSO P-2457, February 1993.

[2]Together, 6.3A and 6.3B form 6.3, which is called by a different name, Advanced Technology.

[3]For example, Program 1, Strategic Forces, has, traditionally at least, funded work associated with SSBN security and survivability. Some of that work properly falls into the RD&T category.

between categories 6.3A and 6.3B, we do not cut off RD&T at that point. It is our understanding and experience that a significant amount of important work goes on in 6.3B that is more properly called RD&T than EP&S. We exclude from the RD&T rubric those activities that are supported by 6.4 (Engineering Development) funds. They are in the EP&S category.

Specific RD&T organizations also are generally outside the scope of our work. However, to lend some concreteness to our discussion, we summarize DON's organizational structure for RD&T. A large part of in-house Naval RD&T activity occurs in five organizations: the four warfare centers—the Naval Air, Surface, and Undersea Warfare Centers (NAWC, NSWC, NUWC) and the Naval Command, Control and Ocean Surveillance Center (NCCOSC)—and the Naval Research Laboratory (NRL). Also closely connected is the infrastructure in the three systems commands—the Naval Air and Sea Systems Commands (NAVAIR and NAVSEA[4]) and the Space and Naval Warfare Systems Command (SPAWAR)—and in the Office of the Chief of Naval Research (OCNR).

Three points regarding this organizational structure are worth noting. First, some of the funding for all but one of the warfare centers and NRL—e.g., about one-fifth for NCCOSC and about one-third for NRL—comes from outside the Navy. Not surprisingly, the exception is NUWC, which receives virtually all its funding from the Navy—since this center deals with undersea warfare. Second, RD&T is not a dominant component of the four centers' budgets; it ranges from about 9 to about 20 percent. However, RD&T accounts for about 70 percent of NRL's budget. Finally, a significant part of DON's RD&T capability resides in its contractors, although we do not know how much. The number is not really important for our purposes.[5] What is important is that *in presenting our framework and discussing our application of it, we assume that out-of-house Naval RD&T capabilities fall within the framework's scope.*[6] Indeed, in Part II of this report, we argue for a flexible view of in-house versus out-of-house capabilities.

SOURCES OF LISTS

The following are possible sources of information for lists of Naval RD&T capabilities:

---

[4]For example, there is considerable RD&T activity in SEA 08, the Nuclear Propulsion Directorate.

[5]We take the view that priority-setting for RD&T should be done across the aggregate (in-house and out-of-house together) of Naval capabilities. Once priorities are set in the aggregate, the issue of "where to buy it," the subject of Part II of this report, can be addressed.

[6]Also within its scope is funding that OCNR allocates to universities and the like via its Science Directorate, which seems mostly to be in the 6.1 Budget Category.

- The five organizations mentioned above issue *Management Briefs*.[7] In examining the briefs, we found that, except for the brief for NRL, they characterize the RD&T capabilities of the organizations by way of mission and subject-matter leadership areas, a manner we judged not directly relevant to our task.[8] In our discussions with high-level representatives of two of the four warfare centers, they agreed with our judgment on this point. (Unfortunately, the short term of the study did not permit interaction with these five organizations beyond those two discussions.)

- Official Navy budget documents listing all the Program Elements. Again, limits on our time and resources, coupled with the complexity of the documents, permitted only minimal use of this source.

- *DEFENSE Program Service*, a reference document issued by Carroll Publishing Company.[9] This document contains a compendium of Navy Program Element Numbers (PENs) and subordinated projects. While we found this document useful for the perspective it gave, it also did not directly characterize DON RD&T capabilities in a way conducive to composing an appropriate list.

- The list of anticipated major Naval acquisition programs provided by the Deputy for Resources, Analysis and Policy (OASN/RD&A/RA&P). This list contained a large number of entries, but it represented a useful starting point and helped us characterize entries on our final list of capabilities, because it permitted us to link those entries to specific programs. In applying our framework, that link facilitated assessing possible effects of different sets of funding priorities (expressed in terms of RD&T capabilities) on those programs.

- Reports of the Tri-Service S&T Reliance Program.[10] These were also useful sources of information about the structure and nature of parts of Naval RD&T per se, as well as in relation to RD&T for the remainder of DoD. As such, they stimulated wide-ranging discussions.

In constructing our list of RD&T capabilities (see Appendix A), then, we drew principally from the Tri-Service S&T Reliance reports, the list of anticipated

---

[7]See the Bibliography for specific listings.

[8]This lack of relevance was not immediately clear. Indeed, as alluded to above and discussed below, part of the problem we faced was to decide how to characterize the RD&T infrastructure. Along with virtually everything else, the way we characterized that infrastructure changed as the study progressed.

[9]See Carroll Publishing Company, *DEFENSE Program Service*, Binders 1 and 2, Washington, D.C., November 1993.

[10]See, e.g., U.S. Department of Defense, Joint Directors of Laboratories, *Tri-Service Science & Technology Reliance: Annual Report*, Washington, D.C., December 1992.

major Naval acquisition programs, and the NRL *Management Brief.* We also drew heavily, of course, on our own knowledge and experience (see Appendix B).

RAND'S FINAL LIST

Our final list of Naval RD&T capabilities comprises 53 entries, arranged alphabetically in Table 2.1. It is not intended to be studied in detail at this point. We present it here for the concreteness needed to facilitate communication of concepts and ideas as we proceed through the discussion of our priority-setting framework, to carry on the application of our methodology through the priority-setting stages (through Chapter Four), and to illustrate some of the choices DON would have to face in making its own list. It is discussed as we move along.

Table 2.1

Alphabetical List of Naval RD&T Capabilities

Air-Vehicle Signature Control and Management	Maintainability, Reliability, Survivability
Amphibious Vehicles	Manpower and Personnel
Anti-air Weapons	Mapping, Geodesy, and Weather
Anti-ship Weapons	Marine Mammals
Ballistic-Missile Defense	Materials
Ballistic Missiles	Mine Warfare
Biotechnology	Naval Oceanography
Carrier-Unique Aspects of Fixed-Wing [F/W] Aircraft	Navigation
Chemical/Biological Defense	Navy Medical
Civil Engineering	Navy Sensors and Signal Processing
Combatant Ships	Non-Maritime Sensors and Signal Processing
Combat Engineering	Nuclear Power/Propulsion
Communications	Nuclear-Weapons Effects
Computer Science & Technology	Precision Strike
Conventional-Weapons Effects	Rotary-Wing Aircraft
Directed-Energy Weapons	Simulation and Modeling
Electronic & Photonic Devices	Space Systems
Electronic Warfare	STOVL Aircraft
Environmental Quality	Submarine Communications
Fixed-Wing Aircraft	Submarine Technology
Flexible Manufacturing	Support Ships
Fuels and Lubricants	Training
Ground Vehicles	Undersea Weapons Technology
High-Performance Offensive Guns	Unmanned Air Vehicles (UAV)
Human Factors	Unmanned Ground Vehicles (UGV)
Information Management	Unmanned Undersea Vehicles (UUV)
	Water-Vehicle Signature Control

NOTE: This list is not official in any Navy or government sense. Rather, it is our characterization of the Naval RD&T infrastructure created specifically for use in applying our priority-setting framework in this report.

One important choice, for example, is roughly how long the list should be. Another is the dimensions or characterizations of Naval RD&T lines. We were not able to make these choices independently[11] or sequentially, but we nevertheless discuss them in order.

Choice of the Number of Entries

Over the course of the study, the number of entries on the list varied from as few as 16 to over 300. A short list characterizes the RD&T work broadly; such broadness is often viewed as appropriate for priority-setting at high levels. A long list permits more precise definition of categories and reduces the likelihood that the rank of a category will reflect a dubious compromise among the rankings of diverse subcategories. Without pretending to have discovered an optimal length, we believe our list represents a workable medium appropriate to the highest Naval levels, at which actual priority-setting must be done.

A key to successful application of the framework by the Navy will be to carry out final list-making (and ranking and priority-setting as well) at a high enough level that participants' inputs and votes are driven primarily by interests no narrower than the Navy as a whole. Otherwise, as is easy to imagine, intra-Navy competition would result in categories centered around favored projects. Selective use of outside consultants and retired high-level military and civilian Naval personnel could be useful for input of expert information.

Characterization of List Entries

Deciding how to characterize the dimensions used to categorize the full spectrum of RD&T activities into an appropriate set of RD&T lines or capabilities was a complex task. Should they be expressed in terms of capabilities that are operationally relevant in the system they support, or weapon systems to which they contribute, or disciplines with which they are affiliated? As discussed below, and as is apparent from our list, we did not choose according to any one of these dimensions. Rather, the list includes branches or sub-branches of science (**Naval Oceanography**), weapons (**Ballistic Missiles**), missionlike categories (**Ballistic-Missile Defense**), processes (**Flexible Manufacturing**), and combat platforms (**Fixed-Wing Aircraft**), among other things.

---

[11]Indeed, not only were these choices iterative, but, as noted previously, the ranking process itself was also involved in the iterations.

We settled on this "apples-and-oranges" approach[12] because it reflects the context within which real-world funding decisions are made. Priority-setters do not simply choose between one high-technology product and another, but, for example, between *improvements* in submarine platforms and in electronic warfare, or between *research* on manpower and on precision strike. The scheme also permits comprehensiveness without requiring exhaustive, probably pointless, listing of all possible categories of RD&T at a consistent level. We have tried to minimize the overlap inherent in an inconsistent categorization scheme (which ours is) by considering under each heading those salient, defining aspects that are not captured elsewhere (see the discussion at the end of Chapter Four). For example, we can single out electronic devices for special attention without having to categorize every other kind of device that might be used on fixed-wing aircraft, because we have a category for the latter. And we do not have to embody in the components that make up the aircraft the full worth of RD&T related to fixed-wing aircraft.

As noted in the opening paragraph of the chapter, constructing the list was not trivial. It was a difficult and time-consuming task that did not end once we progressed—for the first of a number of times—to establishing and applying criteria for priority-setting. When we realized that the priorities for certain capabilities did not make sense, we revisited the list and redefined categories. (For example, parts of fixed-wing aircraft have a different priority from fixed-wing aircraft per se.) As a result, some lines of RD&T were classified in categories that may not always strike some readers as the most obvious. For example, research on materials that might be associated with stealthiness of a future Naval combat aircraft (e.g., in the JAST Program) is under **Air-Vehicle Signature Control and Management,** not under **Materials**.

To give some sense of the path traveled to arrive at the list given in Table 2.1, Appendix A presents and briefly discusses the longest list we devised. Individual entries in the longest list are arranged in groups under the headings given in the table, so the final aggregation is readily apparent. Deliberately, the list in Table 2.1 includes no programs per se; however, as the appendix shows, all anticipated major Naval acquisition programs we know about are included under at least one of the 53 main, or aggregating, entries in the longest list.

The aggregation given in Appendix A is also important as an aid to understanding just what we mean to be included under the various entries in Table 2.1. If not clear at this point, the need for the information in Appendix A will be readily apparent by the end of Chapter Four.

---

[12]The first National Critical Technologies Panel took a similar approach in its 1991 report (see *Report of the National Critical Technologies Panel* in the Bibliography).

Chapter Three

In this chapter, we discuss the two criteria by which we rank, in Chapter Four, the 53 entries in the final list (see Table 2.1).

We sought two important, independent criteria that could be plotted against each other on orthogonal scales, so that dispersions and areas of the plot where groupings that resulted on the two criteria scales might reveal insights into setting overall support priorities. One of the criteria clearly needed to represent the value of the RD&T to the Navy. A criterion of this nature would probably have sufficed for determining support priority in times of relatively unlimited financial resources, as the situation tended to be during the Cold War years. However, in the current drawdown environment, it is important to also somehow characterize possible redundancy within, or even duplication of, Naval RD&T lines or capabilities outside the Navy. All other things being equal, the more likely it is that an equivalent or nearly equivalent capability exists elsewhere *and* might be independently supported anyway, the less is the need for Naval support. We now discuss in some detail our specific embodiment of these two types of criteria.

RELATIVE VALUE

Our first criterion is *relative value of future RD&T to the Navy*. We first explain the qualifiers on that value.

Qualifiers

1st Qualifier:	*Relative* value
2nd Qualifier:	Value of *RD&T*
3rd Qualifier:	*Future* value
4th Qualifier:	Value *to the Navy*

First, the value is *relative*. Value is sometimes expressed in terms of "criticality." But calling a technology "critical"[1] may be taken to mean that support for such a technology is *necessary*, whereas support for noncritical technologies is optional. We concluded that we could not really draw a line above which would lie RD&T categories that DON *must* pursue and below which would lie categories that could be ignored if necessary. DON may be in a position to draw such a line, but we feel more comfortable in ascribing mutually *relative* values to the categories we have selected, for four reasons:

- Assigning value to *all* categories instead of criticality to some further emphasizes our view that all RD&T currently under way is important.

- A critical/noncritical dichotomy does not fit well into the budgetary exigencies that drive priority-setting. For example, if more categories are judged critical than can be adequately funded by the money available, some part of them will not be undertaken, "critical" or not. Additionally, elements deemed "critical" can invariably be worked around, albeit at some (perhaps large) price, dollar or otherwise.

- Certain important RD&T characteristics, e.g., potential for enabling revolutionary operational improvements, cannot be easily captured in a dichotomous critical/noncritical evaluation. (See the following subsection, "Contributing Factors," for a list of these characteristics.)

- We could not agree on a definition of the word *critical*. For example, critical to what, some future mission, or to the very identity of the Navy?

The second qualifier is that, we are, of course, judging the value of *RD&T*. Thus, for example, when we suggest a value for submarine technology, we are not evaluating the worth of past, current, or programmed submarines as a part of the Navy's operational forces, but the worth of *RD&T related to submarines*. Here, relative worth depends partly on the anticipated future value of submarines but also depends partly on the current state of the art of submarines. In our assessments, the time frame for value to accrue is roughly the next 20 to 30 years (see "A Note on the Time Frame and Basic Research" at the end of this section).

That gets us to the third qualifier: Our criterion is based on the value of *future* RD&T. Again using the submarine example, we do not rank per se the value of submarine technology to date, but, given what we already have, the value of *further* RD&T related to submarines. On the other hand, as noted in the

---

[1]Indeed, the term is part of the title of the 1991 *Report of National Critical Technologies Panel* (see Bibliography).

preceding paragraph, we clearly need to understand and judge the current value to make judgments about the additional contribution of future RD&T.

Finally, the relative-value criterion measures value *to the Navy*, i.e., the extent to which a line of RD&T would be expected to contribute to the accomplishment of Naval roles and missions, primarily in the six joint mission areas defined in *Force 2001*:[2]

- Strike, or the ability to project power at any place and time, which depends on various Naval weapon systems, e.g., smart munitions delivered by fighter/attack planes operating from aircraft carriers, cruise missiles launched from submarines and surface ships, and shipborne guns.

- Littoral warfare, or the ability to mass forces and deliver them ashore. Success at this mission relies on amphibious systems and anti-submarine, anti-surface-ship, and mine-warfare elements.

- Surveillance, which requires sensors and means to process and transmit sensor data. It includes, for example, the use of netted sensors to help air, land, and sea platforms with targeting.

- Strategic deterrence. Arms control agreements have shifted a greater burden onto the submarine-borne leg of the U.S. Strategic Triad.

- Space and electronic warfare, which is intended to exploit the electromagnetic spectrum while denying its reliable use by an enemy. It involves technologies that enhance signals management, operational deception, countersurveillance, and electronic combat.

- Strategic sealift, which includes afloat prepositioning; seaborne movement of surge-unit equipment and sustaining supplies, e.g., via fast supply ships; and protection of assets carrying out these functions, e.g., through convoying operations.

RD&T can also contribute to the readiness and support functions underlying these missions, for example, through more capable systems for depot- and intermediate-level maintenance.

The relative-value scale we devised spans four main levels, designated A (highest) through D. However, below, three boundary cases are introduced that, in effect, are additional levels, and result in a total of seven levels.

---

[2]See U.S. Navy, Chief of Naval Operations, 1993.

Contributing Factors

In judging relative value to the Navy, we took into account several factors not included explicitly above,[3] although all of them are in some way related to success in the mission areas:

Potential for Revolutionary Capability Improvements. We judge RD&T categories to be of greater value if the capability gains realized from them may go beyond increasing the efficiency and effectiveness of *current* operational and tactical practices. By "revolutionary improvements" we mean improvements that offer a whole new way of doing business in a mission area. For instance, "precision" strike, accomplished by the Tomahawk and other cruise missiles, allows fixed enemy targets to be destroyed from international waters with minimal friendly and civilian casualties and collateral damage, and without use of ballistic missiles, which we judge to be politically unacceptable. Before the precision-strike era, this objective, so constrained, would have been impossible to meet.[4]

Speed of Development. How long will it take the technology in question to enter full-scale engineering development? All other things being equal, the sooner an investment in a given RD&T category will generate operational payoffs, the more valuable it is. Not only are long-term payoffs more heavily discounted, they are more uncertain. However, all other things are rarely if ever equal, and the prospect of revolutionary gains 20 to 30 years into the future may render current investment more attractive than it would otherwise be.

Avoiding Technological Surprise. We tended to assign higher value ratings to capabilities in which advances by a potential adversary could put the United States at a distinct military disadvantage if those advances caught the United States unawares. To a great extent, this factor is part of the preceding two factors and is not solely within DON's purview. Responsibility for avoiding technological surprise is shared with both the intelligence community and the private sector. But it still played a role in our deliberations, and we wanted to emphasize this by listing it separately here.

Technological Risk. All other things being equal, if an RD&T category is associated with what we judge to be high technological risk, then it is downgraded accordingly. For example, we judge that high-power directed-energy (beam)

---

[3]One factor not included in "relative value" is the infrastructure per se. That factor is, however, tacitly included in our other criterion, breadth of demand.

[4]Our deliberations also included recognition of the important role played by intelligence (for weapon and mission planning) in enabling precision strike—to varying degrees, of course, depending on the specifics.

weapons would probably not achieve *Naval* operational capability in the next 20 to 30 years, even with reasonably high levels of funding.

Potential for Multiple Applications. If an RD&T category offers the promise of advancing Naval capabilities in more than one area, then, all other things being equal, we judge its value to be higher than that of another category engendering advances across a narrower front. For example, while directed-energy weapons are often thought of in the context of theater missile defense, they could also enhance fleet air defense, improve the Navy's airborne anti-ship capability, and provide an antisatellite capability. RD&T related to support ships, on the other hand, would have more limited applications. Some applications have inherently greater value than others, so correspondingly heavier weighting must be used.

Responsiveness to Evolving Threat. How important is a given RD&T category for operations against the types of adversaries the Navy and Marines must be prepared to fight in the future? During the Cold War, DON prepared to conduct open-ocean operations against large numbers of nuclear submarines and sophisticated surface ships. Now, the challenge is expected to come from regional powers possessing a handful of high-performance fighter/bombers, a few diesel attack submarines, many inaccurate ballistic missiles, some GPS-guided cruise missiles, chemical/biological weapons, and, perhaps, a few crude nuclear devices. Technologies need to be evaluated on the basis of their appropriateness for use against such opponents. More specifically, it will be possible to make trade-offs that could not be made during the Cold War. For example, DON might be able to back off on its push for increased submarine stealth in favor of concentrating on technologies that would likely be of greater value in littoral warfare.

We note that these factors are not independent of one another, so it is not possible to view any of them in isolation. Also, most of the RD&T categories we are considering would not score high on all these factors; therefore, trade-offs must be made to arrive at a composite value. Finally, we are not assessing which RD&T areas would promote *new* force postures or roles and missions for DON. As stated in Chapter One, we are assuming the postures/missions as given by DoD's "Bottom-Up Review" briefing and various official defense policy guidelines.

A Note on the Time Frame and Basic Research

Use of a 20-to-30-year period for contributions is increasingly prejudicial as the R&D category moves from 6.3 (Advanced Development) to 6.1 (Research), because corresponding technology maturities are farther and farther out in the future. For 6.1, they are probably beyond that time frame in many cases. It should not be surprising, then, that we do not rank basic research very high in relative value. It is inherently difficult to know when, where, or how an investment in basic research will pay off, even though it is certain that at least some of it will pay off **some**time, **some**where, **some**how. Furthermore, basic research considered as a whole is required to maintain technological superiority over the long term. We briefly return to this important topic in Chapter Four.

BREADTH OF DEMAND

If choices have to be made between pursuing two categories or lines of RD&T that are of similar value, DON decisionmakers might choose the one that depends most heavily on their support. If one of the other services is carrying out RD&T efforts in a category, or if another government agency or the private sector is, and it appears likely that support will continue, DON may decide to reduce its support or to bow out. Of course, the results of these RD&T efforts may not be as suitable for Naval purposes as they would have been had DON carried out the research itself. However, they would be better than nothing, which would be the yield if DON pulled out when it was the sole supporter. Thus, it makes sense for DON to consider reducing or eliminating support for RD&T in categories in which it knows that results of at least *some* relevance will continue to emerge and an active RD&T infrastructure will be sustained until it becomes desirable to resume Navy funding.

There are, of course, good reasons for redundancy or duplication; e.g., each can foster competition and provide a hedge against fewer entities not producing what is in demand, at least not as soon or as well as might otherwise be the case. But in financially tighter and less-threatening times,[5] it seems prudent to attempt to identify potential redundancy and at least reduce it to a more financially efficient level.

The prospects for *continuation* of external support are important. Lines of RD&T are difficult to restart once terminated. We use the likely future breadth of demand within the marketplace as a proxy for the prospects that others will continue RD&T activities in a given category.[6] The breadth of demand for an RD&T category is expressed in terms of one of four classes:

---

[5]Again, times are less threatening because of the Cold War's end, which means that the very existence of the United States is less threatened.

[6]Note that, generally, the broader the demand is, the more certain it is that there will be a supply; but the converse is not as clear, which is the reason we chose breadth of demand as our second criterion rather than some more-traditional measure of supply, such as number of suppliers or availability.

- Class 1, Navy-unique: Some Naval RD&T is likely to be of interest only to DON, and thus the department will have to support it if it is judged to be of sufficient value to merit continuation. The **Nuclear Power/Propulsion** category is an example of a Navy-unique RD&T category. Civilian nuclear-power R&D is essentially moribund; therefore, if DON wants to benefit from further R&D in this category, it will have to support the activity itself. The same applies to R&D on weapon systems specific to the Naval context. For example, no one outside the Navy is likely to sponsor all needed RD&T related to torpedoes.[7]

- Class 2, Other DoD: Some lines of RD&T that interest DON are being pursued by the other services or by such defense agencies as the Advanced Research Projects Agency (ARPA) and the Defense Nuclear Agency (DNA). While that research will probably not completely satisfy DON's specific goals, it would be easier to adapt and extend than a line on which all relevant activity had stopped because of a cessation of DON funding. For example, DON might be able to realize considerable utility in ongoing and future Air Force RD&T on the non-carrier-unique aspects of fixed-wing combat aircraft or air-to-air weapons.[8] If DON had to choose between continuing RD&T on ship self-defense weapons and air-to-air weapons, it would doubtless consider that much relevant work on the latter would continue without its support.

- Class 3, Other government: In a few categories, RD&T is being carried out not only by the military but also by government agencies outside DoD, such as the National Aeronautics and Space Administration (NASA). Examples include research on unmanned aerial vehicles (UAVs), communications satellites, and human factors.[9] Because of the broader demand for RD&T in these categories, DON could suspend or reduce funding for them with more confidence that results of some use to DON would continue to emerge (than it could in the case of RD&T of military interest only).[10]

- Class 4, Private-sector and foreign interest: In some areas, DON could benefit from RD&T activities funded by the U.S. private sector or by foreign governments or commercial sectors. Fields such as computer science and biotechnology have broad application, and a great deal of diverse research

---

[7]Other nations might develop torpedoes, but even if such torpedoes could and would meet the United States' direct needs, there are probably broader national security and national policy considerations that would drive the United States to do at least some of its own RD&T on torpedoes.

[8]A Navy-unique aspect of air-to-air weapons might be associated with radar sea-clutter.

[9]Of course, human-factors research is being carried out in the civil sector, too, but little of it is considered to be relevant to the military.

[10]We also note that some human-factors research is unique to the Navy, e.g., that associated with life aboard combatant ships and submarines.

will continue in these areas without DON support or with less of it. We do not include as part of foreign breadth of demand any demand for items that the DoD[11] would generally not be willing to buy or sell. For instance, there may be a broad demand for various nuclear technologies among developing countries, but that is not a demand the DoD would want to meet.

This scale captures the relationship between increasing breadth of demand (moving from Class 1 to Class 4) and the decreasing riskiness of allowing others to carry out Navy-desired RD&T activities. Where demand is limited to the government, and especially to DoD, risk is higher—one obvious reason being that other DoD departments and agencies may perform an analysis similar to that suggested here and conclude that they can discontinue some types of RD&T because DON is pursuing them. Whatever the reason, once DON cedes responsibility for research, it no longer has control over continuation of that research, although it may have a window of opportunity to pick up funding when another agency drops out.

At some point, researchers will move on to other interests, and restarting a line of RD&T could be very time-consuming and expensive. And it may be impossible to achieve newly required or desired operational capabilities by target dates. Thus, the breadth-of-demand criterion needs to be applied with as much care as the relative-value criterion.

As to the independence of the criteria for use on an orthogonal (X-Y) plot, we consider breadth of demand to be independent of relative value as we use the terms, because there is a full range of breadth-of-demand classes that are of highest relative value and vice versa (see Chapter Four). We now show how RD&T capabilities can be plotted along axes denominated by the two criteria.

---

[11]Clearly, the State and Commerce Departments may have different perspectives; we do not make judgments on those perspectives.

RANKINGS AND SUPPORT PRIORITIES

We begin this chapter by describing and discussing the ranking process we used. Bear in mind, however, that as noted at both the beginning and end of Chapter Two, the list-making and ranking processes were highly intertwined and the whole framework-application process was highly iterative. The remainder of the chapter covers the process of going from the rankings by each of the two criteria to the support-priorities framework, gives our final rankings in the framework, and closes with a discussion of some of the rankings.

RANKING PROCESS AND FINAL LISTS

Rankings were carried out using a sequence of a type of Delphi[1] process over a period of about four weeks.[2] *Delphi*, as we use the term, is a method of building consensus among panels of specialists by using a number of rounds of "blind" (i.e., private, individual) ratings or rankings and rounds of open discussion of differences, then reevaluation. The group that did the Delphi comprised eight of the report's ten authors (see Appendix B). Early Delphi rounds were carried out blind; later rounds, including the final, full-day session, were open. In discussions of blind-round results in open meetings, we attempted to

---

[1]See, e.g., Norman C. Dalkey, *The Delphi Method: An Experimental Study of Group Opinion*, Santa Monica, Calif.: RAND, RM-5888-PR, June 1969.

[2]We had originally intended to give each RD&T category a value score from 1 to 10 for each of the six mission areas listed in Chapter Three. These scores would have been based on the results of studies and campaign analyses of future contingencies, as well as on expert opinion, and would have taken into account the contributory factors described in Chapter Three. The scores were then to have been weighted and summed to give a final "value," and the values were to have been grouped along some scale such as the A-to-D scale we eventually devised.

It soon became apparent that this plan would not work, for at least two reasons. First, and probably most important, this scoring would have implied more quantitative certainty and a finer gradation of values than we could really achieve—indeed, than may be achievable in principle. Second, time and resources were not available to even attempt such a comprehensive, systematic assessment. There was not even enough time to be sure of doing all the bookkeeping correctly, much less to carry out separate studies and joint campaign analyses. Indeed, it is unclear whether appropriate models of such campaigns even exist.

rationalize the sometimes-wide variability in outcomes across the group, particularly in the relative-value outcomes. The open-session rankings were iterative, but each RD&T category was done simultaneously by both criteria in each iteration. Often during an open session, especially in the earlier sessions, the group would break out of the process of ranking one category to revisit another category's rankings. Sometimes the revisit would involve deleting, adding, or redefining categories.

Detailed Description of Process

In more detail, in resolving disagreements among rankers for a particular RD&T category, we often found that one or more rankers had temporarily lost sight of one or more of the four qualifiers on relative value given at the beginning of Chapter Three: (1) relative, *not* absolute, value, (2) RD&T, not current or programmed force structure, (3) future, not current or past, RD&T, and (4) value to the Navy in, primarily, its joint mission areas. As a result, we made reiteration of these four qualifiers a virtual ritual early on as we took up each RD&T category.

Similarly, sight was often lost of the things that were to be included or not included in an RD&T category, so we would repeatedly go over the items in that category (under that heading in Appendix A)—and, as noted, redefine categories and delete or add to them and others as we moved along (and revisited). Naturally, when a category was added, deleted, or redefined, it was often necessary to move one or more of the various entries under the corresponding heading in Appendix A to another heading. Stated another way, the longest list, once established, remained nearly unchanged, but its aggregation within the RD&T categories (ranging a few in number around the final 53) changed quite often. And, of course, the category headings changed from time to time, as well.

Continuing, with some recapitulation and with concrete examples, to reach consensus, we sometimes split RD&T categories, relative-value levels, and breadth-of-demand classes. For example, at times, a member of the group was willing to change a ranking if one of the RD&T categories was narrowed and the excluded subcapabilities were listed in a new category or were combined with another. Sometimes, the whole group would judge that a change was necessary to clarify the meaning of an entry.[3] For example, there was initially one category for fighter/attack aircraft, but we ended up with four separate ones, **Fixed-Wing Aircraft**, **Air-Vehicle Signature Control and Management**, **STOVL Aircraft**, and

---

[3]We found the tasks of defining and ranking categories so intertwined that we recommend that, in any subsequent application of this methodology, the group doing the ranking have the capability to change the list.

Carrier-Unique Aspects of Fixed-Wing Aircraft. As indicated, the best way to come to an understanding of the meaning we attached to each of the 53 categories is to examine Appendix A, which shows the longest list in the form of its aggregation to the final 53. For instance, it shows the JAST Program, associated with a next-generation Naval fighter/attack aircraft, under all four of the above categories.

At one point, to accommodate disagreement, we found it necessary to create a borderline class on the breadth-of-demand scale. Specifically we could not agree whether one RD&T category, **Amphibious Vehicles**, was Navy-unique or Other DoD. The issue centered on the definition of *amphibious*—on whether surface-effects vehicles have practical land capabilities. As a result, we finally created a separate intermediate class, 1.5, for it. Also, to reach a consensus on some relative-value ratings, it was necessary to recognize cases on all three boundary levels, AB, BC, and CD. As a result, we ended up with seven rather than four levels of relative value, but rather than redesignate them A through G, we retained the basic four and the borderline cases.

Participants in Decisionmaking

To carry out a process such as the one just described, the participants clearly must have not only a good understanding of the Navy, its current capabilities and anticipated future needs, and its opportunities in a joint world, but they must also have as top priority the interests of the Navy as a whole rather than any individual part or parts of it. Thus, if the Navy decides to try to use the framework, it will, in the end, need to do so at an appropriately high level, e.g., on the military side, above the N86/N87/N88 level,[4] a level at which the leaders necessarily have the interests of the corresponding platform types foremost in their minds.

Relative-Value List

The final version of the relative-value list (i.e., our final ranking) is given in Table 4.1. Several features of this list need emphasis or re-emphasis. First, the three borderline cases, AB, BC, and CD, are shown in boldface to highlight, and only to highlight, the boundaries. No other meaning is intended by the bold-face type in the table (in the text, on the other hand, *all* RD&T categories

---

[4]This level is the highest one specifically responsible for the three main categories of combatant platforms, surface ships (N86), submarines (N87), and aircraft (N88).

Table 4.1

Prioritized Naval RD&T Capabilities, by Relative Value

RD&T Capability	Relative Value	Breadth of Demand
Mine Warfare	A	1
Navy Sensors and Signal Processing	A	1
Air-Vehicle Signature Control and Management	A	2
Anti-air Weapons	A	2
Ballistic-Missile Defense	A	2
Non-Maritime Sensors and Signal Processing	A	2
Precision Strike	A	2
STOVL Aircraft	A	2
Space Systems	A	3
Unmanned Air Vehicles (UAV)	A	3
Electronic & Photonic Devices	A	4
Navigation	A	4
Combat Engineering	**AB**	**2**
Electronic Warfare	**AB**	**2**
Carrier-Unique Aspects of Fixed-Wing Aircraft	B	1
Nuclear Power/Propulsion	B	1
Unmanned Undersea Vehicles (UUV)	B	1
Water-Vehicle Signature Control	B	1
Ballistic Missiles	B	2
Chemical/Biological Defense	B	2
High-Performance Offensive Guns	B	2
Communications	B	4
Computer Science & Technology	B	4
Flexible Manufacturing	B	4
Information Management	B	4
Rotary-Wing Aircraft	B	4
Support Ships	B	4
Combatant Ships	**BC**	**1**
Maintainability, Reliability, Survivability	**BC**	**1**
Submarine Technology	**BC**	**1**
Fixed-Wing Aircraft	**BC**	**2**
Anti-ship Weapons	C	1
Naval Oceanography	C	1
Submarine Communications	C	1
Undersea Weapons Technology	C	1
Amphibious Vehicles	C	1.5
Ground Vehicles	C	2
Manpower and Personnel	C	2
Conventional-Weapons Effects	C	3
Nuclear-Weapons Effects	C	3
Materials	C	4
Simulation and Modeling	C	4
Unmanned Ground Vehicles (UGV)	C	4
Training	**CD**	**4**
Marine Mammals	D	1
Navy Medical	D	1
Directed-Energy Weapons	D	2
Human Factors	D	3
Mapping, Geodesy, and Weather	D	3
Biotechnology	D	4
Civil Engineering	D	4
Environmental Quality	D	4
Fuels and Lubricants	D	4

NOTE: Borderline categories are in boldface to indicate boundaries.

are boldfaced). Within any of the seven groups having the same relative-value level, no distinction in value is intended. Rather, the second-level ordering in such groups is by breadth of demand. Similarly, there is a range of classes of breadth of demand within most of the levels of relative value. For example, the highest value level, A, has entries from all four of the original classes of breadth of demand. The one borderline, breadth-of-demand case, class 1.5, **Amphibious Vehicles**, stands out clearly in the right column of the table, so it is not shown boldfaced. Ordering is alphabetical in subgroups for which both criteria have the same value (*level* for relative value and *class* for breadth of demand), which we call *bins* or *cells*.

Breadth-of-Demand List

Table 4.2 shows the same information as Table 4.1, but sorted with breadth of demand as the primary criterion. Within groups of equal breadth of demand, the ordering is by relative value. As with Table 4.1, within cells, the ordering is alphabetical. Note that the boldface-type rows of the same relative-value level from the preceding table are no longer grouped together, so they do not represent, as a group, the same kind of overall boundary. However, within each class of breadth of demand they still have the same meaning.

GENERIC ORTHOGONAL PLOT

Neither of the tables furnishes an easy visual presentation of the ranking results. Figure 4.1 shows relative value on the ordinate and breadth of demand on the abscissa. It also shows 53 small circles, which give, without identification, the cell locations or final rankings of each of the 53 RD&T categories. The names are omitted for now to facilitate focusing on just the framework. Again, all RD&T categories in a cell (including the separate borderline cells), have identical ranking. The locations of circles within individual cells are of no significance. Likewise, there is no significance to the size or shape of the cells— larger cells are needed to hold more circles.

FROM RANKINGS TO SUPPORT PRIORITIES

As noted in Chapter Three, the main motivation for using independent criteria is to enable creation of a more meaningful *X-Y* plot so that insight can be gained into how the two criteria together can be used to indicate support priorities. As noted there also, in relatively affluent times, the things that would deserve the highest support appear to be those judged of highest value. However, with more modest threats abroad and, therefore, leaner times militarily, it seems important to consider adjusting upward the affluent-times support priority of

Table 4.2

Prioritized Naval RD&T Capabilities, by Breadth of Demand

RD&T Capability	Relative Value	Breadth of Demand
Mine Warfare	A	1
Navy Sensors and Signal Processing	A	1
Carrier-Unique Aspects of Fixed-Wing Aircraft	B	1
Nuclear Power/Propulsion	B	1
Unmanned Undersea Vehicles (UUV)	B	1
Water-Vehicle Signature Control	B	1
Combatant Ships	**BC**	**1**
Maintainability, Reliability, Survivability	**BC**	**1**
Submarine Technology	**BC**	**1**
Anti-ship Weapons	C	1
Naval Oceanography	C	1
Submarine Communications	C	1
Undersea Weapons Technology	C	1
Marine Mammals	D	1
Navy Medical	D	1
Amphibious Vehicles	C	1.5
Air-Vehicle Signature Control and Management	A	2
Anti-air Weapons	A	2
Ballistic-Missile Defense	A	2
Non-Maritime Sensors and Signal Processing	A	2
Precision Strike	A	2
STOVL Aircraft	A	2
Combat Engineering	**AB**	**2**
Electronic Warfare	**AB**	**2**
Ballistic Missiles	B	2
Chemical/Biological Defense	B	2
High-Performance Offensive Guns	B	2
Fixed-Wing Aircraft	**BC**	**2**
Ground Vehicles	C	2
Manpower and Personnel	C	2
Directed-Energy Weapons	D	2
Space Systems	A	3
Unmanned Air Vehicles (UAV)	A	3
Conventional-Weapons Effects	C	3
Nuclear-Weapons Effects	C	3
Human Factors	D	3
Mapping, Geodesy, and Weather	D	3
Electronic & Photonic Devices	A	4
Navigation	A	4
Communications	B	4
Computer Science & Technology	B	4
Flexible Manufacturing	B	4
Information Management	B	4
Rotary-Wing Aircraft	B	4
Support Ships	B	4
Materials	C	4
Simulation and Modeling	C	4
Unmanned Ground Vehicles (UGV)	C	4
Training	**CD**	**4**
Biotechnology	D	4
Civil Engineering	D	4
Environmental Quality	D	4
Fuels and Lubricants	D	4

NOTE: Borderline relative-value categories are in boldface but are not arranged to indicate boundaries as in Table 4.1.

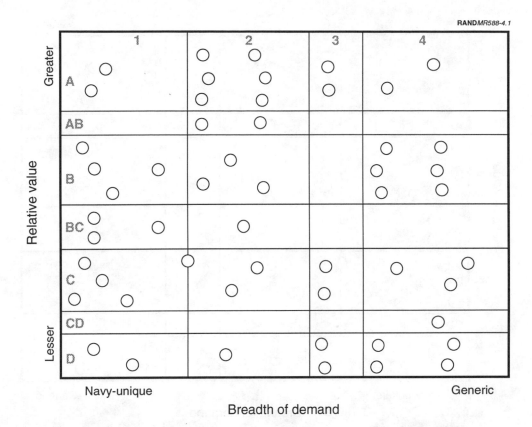

Figure 4.1—Orthogonal Plot for Relative Value and Breadth of Demand

things that, although not deemed as valuable as others, probably would not receive support outside the Navy. More generally, the narrower the breadth of demand is, the greater would be the upward adjustment of support priority.[5]

INTERPRETED GENERIC ORTHOGONAL PLOT

Figure 4.2 illustrates how relative-value by demand-breadth regions can be interpreted to imply support priorities. It shows, *for illustrative purposes only*, a group of shaded cells that would be of higher (although not necessarily equal)

_____

[5]At one point during the course of the study, we considered making a corresponding set of "affluent and more-threatening times" rankings. However, we decided that set would degenerate, for all practical purposes, to a one-dimensional situation from a support-priorities perspective: Breadth-of-demand classes would be the same but, in affluent times, would not be nearly as relevant to support priorities. Continuing the "permeating-atmosphere" discussion from Chapter One, the driver for choosing breadth of demand as a criterion appears to be independent of the BUR permeating atmosphere per se and is, instead, a consequence of the leaner-times environment resulting, in turn, from the end of the Cold War.

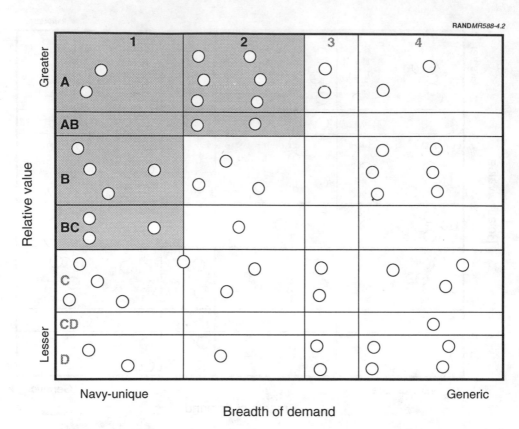

NOTE: Shaded cells indicate higher, but not necessarily equal, Naval support priority. Unshaded cells indicate lower, but not necessarily equal, support priority.

Figure 4.2—Inferring Naval Support Priorities from Orthogonal Plot

support priority than those in the unshaded cells. Within the shaded section, some cells might be considered of higher Naval support priority, e.g., RD&T lines in cell A1 would be singled out as highest. The same kind of reasoning applies to the unshaded parts of the figure, where RD&T lines in the cell at the lower right (D4) would have lower Naval priority than all the others. Generally, then, the farther up and to the left RD&T lines fall in the diagram, the higher the corresponding support priority. Similarly, Naval support becomes less important to the Navy, and thus is of lower priority, as one moves down and to the right on the diagram.[6]

_____

[6]None of the authors has a supporting background in the appropriate parts of what probably falls in the intersection of economics and psychology. But, imagining appropriate continuous linear numerical scales on the diagram's axes, some of us conjectured that iso-support contours might be something like a family of rectangular hyperbolas (with the tails truncated, of course). Others felt that a family of parallel straight lines sloping from lower left to upper right would represent this

CANDIDATE APPROACH TO IMPLEMENTATION

Although beyond our original scope, we observe that a candidate approach to implementation of the indicated support priorities is to fund all, or nearly all, of the good ideas or proposals in both RD&T lines shown in the upper left cell, varying down to funding, say, 10 percent of the good ideas in each RD&T line in the lower right cell. A *good idea* is one that is judged likely to have relevant success by those responsible for allocating funds within an RD&T line or category.[7] Such an implementation approach would probably be highly iterative and could involve adjusting funding levels in each RD&T line or category to accommodate the availability of good ideas. Similarly, the threshold for what is good and the indicated percentages might have to be adjusted, depending on the amount of the total funding available.

FINAL RANKINGS ON ORTHOGONAL PLOT

Figure 4.3 is Figure 4.1 with the circles identified.[8] Rankings and the corresponding list are, as noted, only our "best cut" at applying the framework. In the end, the list and ranking of its entries must be done by the Navy itself. *The most important output in this part of the report is the framework per se*. Again, rankings reflect *relative value of future RD&T to the Navy*. Past and current contributions of RD&T in a category are not directly reflected, and the value of the technology alone and in itself, or as embodied in a system, does not determine the ranking.

Whereas a definitive setting of support priorities using a framework such as this one must await Naval application, several trends are apparent that we believe would remain in an official Naval ranking:

- RD&T categories judged of greatest value tend to be associated, one way or another, with the information revolution—with computers and communications. **STOVL Aircraft** (A2) is a notable exception.

- RD&T associated with the three main types of combatant platforms fall into relative-value class BC: submarines (**Submarine Technology**, BC1), surface ships (**Combatant Ships**, BC1), and **Fixed-Wing Aircraft** (BC2). The main

---

theoretical limiting situation. The shading in Figure 4.2 would seem to support both conjectures and, for that matter, virtually any family of curves that are concave to the lower right.

[7]It might appear that by talking about "good" ideas we are opening up a third criterion dimension for the 53 RD&T categories, but that is not the case. No judgment is intended (in the footnoted sentence) about the goodness of any of the 53 RD&T categories per se. Rather, we refer to the quality and prospects for relevant success of RD&T proposals *within* a category.

[8]In some cases, the wording for categories has been shortened to avoid crowding in the figure, e.g., **Air-Vehicle Signature Control** (cell A2).

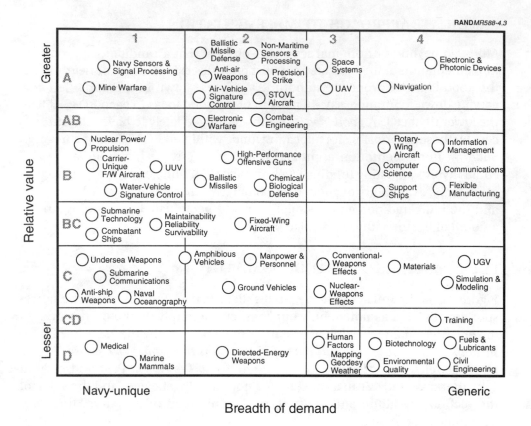

Figure 4.3—Rankings of Naval RD&T Capabilities

reason they are that low is that current platforms are already so good. We judge that over the next 20 to 30 years, significantly larger value will come from things that are installed in, hung on, or communicate with these types of platforms, e.g., sensors, weapons, communications equipment, and, of course, the attendant computers and software, than from advances in the platforms themselves. Note, however, that other platforms, what might be called nonclassical ones, **STOVL Aircraft** (A2), **UAVs** (A3), **UUVs** (B1), and **UGVs** (C4), are RD&T lines in their own right.

- RD&T categories that are clearly associated with general basic research tend to be located toward the lower right, e.g., **Materials** (C4) (with, as noted in Chapter Two, exceptions included in other categories) and **Biotechnology** (D4).

In connection with the last bulleted topic, this priority-setting approach—when viewed from a larger perspective than the Navy's—is clearly suboptimal, because it takes no account of the value of any RD&T category beyond the Navy.

Its use raises an important issue that is outside the scope of this report: If the Navy were to become less supportive of Research (6.1), who should, in the national interest, pick up the slack? One possibility is the Office of the Secretary of Defense, specifically, the Office of the Director of Defense Research and Engineering. Wherever basic research ends up, it is important that its funding be long-term and stable; otherwise, many of the best people will go elsewhere.

DISCUSSION OF SOME SPECIFIC RANKINGS

We did not document the details of the deliberations and arguments that resulted in all the final rankings shown in Figure 4.3. However, to aid readers wishing to obtain more insight into specific rankings, we now briefly discuss a number of rankings, particularly those that seemed most counterintuitive to our many audiences for the corresponding briefing. As noted earlier, to facilitate rationalization of the rankings, it is useful and sometimes, even necessary, to both consult Appendix A and constantly review the four qualifiers on relative value given at the beginning of Chapter Three.

- The relatively low value attributed to **Directed-Energy Weapons** (D2) reflects our judgment of the attendant high risk. If these weapons could be made operationally useful (e.g., on the deck of a ship) at reasonable cost in the indicated time frame, they would probably be of higher or even highest relative value. But we judge that they would not be, even if they were to receive highest support priority.[9]

- The relatively low value of **Simulation and Modeling** RD&T (C4) reflects our understanding of the state of the art: It is very good. The payoff will come from doing more of it, i.e., deftly applying it, and not so much from further RD&T. Also, we note that because **Simulation and Modeling** RD&T is in breadth-of-demand class 4, the potential negative impact on the Navy of mistakenly ranking it too low in value is lessened by the likelihood that any unexpected breakthroughs in this category of RD&T would be done anyway, by others.

- The seemingly low value of **Training** RD&T (CD4) does not suggest that we view training itself as unimportant; indeed, we believe it is of highest importance. As for RD&T in that category, however, we feel it will generally not have as high a payoff as the higher-ranked categories, and, as reflected in its breadth-of-demand class 4, it will be covered elsewhere anyway.

---

[9]We also note that, if we had been applying the framework to the Air Force rather than to the Navy, we might have ranked them higher—reflecting our view of their relatively higher practical potential from space and high-altitude aircraft.

- The highest value for **Mine Warfare** (A1), which also includes mine countermeasures, reflects our judgment of both the importance of this subject in the new world of littoral warfare and the potential of the products of the information revolution to enable great strides. Virtually every regional power has at least a rudimentary mining capability, and large numbers of even primitive mines can be dangerous during high-tempo combat operations. Moreover, the leverage is high in the new jointness world for mine countermeasures RD&T, because of the anticipated overall value of sealift.

- Shallow-water ASW is one subject that is conspicuous by its absence from the list of 53 categories of RD&T. This important topic is included in **Navy Sensors and Signal Processing** (A1). As with **Mine Warfare**, shallow-water ASW is expected to be important in the new world of littoral warfare. However, because of the uncertainty of the importance of adversary submarines in littoral warfare compared with other platforms, we decided not to make shallow-water ASW a separate RD&T category.

- **High-Performance Offensive Guns** are ranked B2 because their potential is deemed important to the future of the Navy. From the demand standpoint, other military services are interested in doing serious research on high-performance field guns. The Air Force, for example, in the course of its missile-defense program, has experimented with hypervelocity guns that could hurl projectiles into space. In terms of relative value, any breakthroughs in this RD&T area would significantly improve the Navy's shore-bombardment capability from existing platforms, making opposed amphibious landings easier. However, we judged the marginal capability increase that could be realized here to be smaller than for those RD&T lines ranked in relative-value level A, because the Navy already has a potent shore-bombardment capability with carrier aircraft and Marine attack helicopters based on amphibious assault ships—and that capability will get better yet.

- RD&T in **Submarine Communications** (C1) is given relative-value level C (fairly low) because we judged the contribution of submarines in the new world to be less than for other platforms, even if communications were to be greatly improved. In contrast to some of the other rankings above, the relatively low ranking is not a reflection of our belief that these communications capabilities are already good enough, particularly in frequency bands that would be useful in joint operations in littoral areas. Again, instead, it reflects our view that submarines will play a relatively less important role in such areas, even with much-improved communications capabilities.

- **Flexible Manufacturing** (B4) is currently a topic of great interest and importance in both the private and public sectors—and is thought by some to be nearly a panacea for many DoD needs. We also think it is important, but do not rank it any higher because we judge the investment required to make it pay off for defense to be much higher than is generally recognized in many Navy (and DoD) circles.

Chapter Five

RECAPITULATION AND CONCLUDING OBSERVATIONS FOR PART I

No matter how it is done, priority-setting is a difficult and iterative process. Our framework for priority-setting comprises four coupled (interdependent) steps: (1) list-making, (2) criteria specification, (3) ranking, and (4) determination of support priorities. The following bulleted items recapitulate important aspects of our four-step framework:

- The list must be long enough to permit a characterization of the Naval RD&T infrastructure that is adequate for, and appropriate to, priority-setting but short enough to be manageable. Our list comprises 53 "apples-and-oranges" categories of RD&T. If the Navy decides to apply the framework itself, it would not surprise us if the resulting characterization were different and the corresponding list a different length.

- The specified criteria must reflect the potential contributions and attributes of the RD&T infrastructure that are most important to the Navy in the expected future environment, which is less threatening and less affluent than during the Cold War. We chose two criteria, *future relative value to the Navy* and *breadth of (friendly) worldwide demand for the products of the lines of Naval RD&T*.

  — Choice of relative value rather than some form of the critical/noncritical dichotomy reflects our view that no line of RD&T is critical, because workarounds invariably exist (at a price, of course), and use of a critical/noncritical demarcation promotes an untenable situation in which there are an unfundably large number of "critical" lines.

    - Use of a 20-to-30-year period for contributions to relative value is increasingly prejudicial as S&T moves from 6.3 (Advanced Development) to 6.1 (Research), because corresponding technology payoffs are less certain and are farther and farther out in the future. For 6.1, many payoffs are probably beyond this time frame. It should not be surprising, then, that we do not rank basic research very high in relative value, because it is inherently difficult to know when, where, or how an investment in basic research will pay off, although it is certain that at least some of it will pay off—**some**time, **some**where, **some**how.

- • Furthermore, basic research in general is required to maintain technological superiority over the long term.

— Breadth of demand addresses the issue, Will the product be available elsewhere anyway in the future? It was chosen over more traditional measures of the underlying concept of availability and supply, because it permits classes to be scaled monotonically together with parts of the public and private sectors.

• Ranking results can be displayed within a matrix or orthogonal plot similar to that in Figure 4.1. It is important that the following constraints be regarded in the ranking:

— The ranking, as well as the other steps (because of the iterative nature of the process overall), must be done at high enough levels of the Navy that the parties involved have the interests of the Navy as a whole as their top priority. It would be inappropriate to have the ranking done at the N86/N87/N88 level, for example, because platform-related interests would appropriately be at the top of these participants' lists of priorities. Also, it might be appropriate, for input of information only, to include outside experts and retired Naval personnel at intermediate stages of ranking.

— The word *future* is important in ranking a line of RD&T, because we are concerned with, for example, ascertaining the value of *further* submarine RD&T—not the future value of submarine RD&T that has already been done.

• We encourage the DON to use the framework offered here as a means of structuring priority-setting at high levels within the department. Short of that, the framework could be used as a "sanity check" or consistency check on priorities arrived at by some other means.

It is essential to be able to change the list, both in number and content, and to be virtually ritualistic about reminding the participants regularly and often during the application of the framework of (1) the meaning of "relative value," by going over the four qualifiers—*relative, future, RD&T,* and *to the Navy*—and (2) the content or meaning of each RD&T category, by referring to the corresponding entry in Appendix A.

We emphasize that we regard the framework itself as the primary output of Part I of the report. Our application of the framework should be viewed only as *our* best cut, intended as much to provoke debate as to meet with agreement. We do not have sufficient Naval expertise to confidently advise DON on what its RD&T priorities should be.

PART II

DEVISING NEW SOURCING STRATEGIES

Chapter Six

GENERAL CONSIDERATIONS FOR STRATEGIC SOURCING

We now turn our attention from setting RD&T support priorities, i.e., deciding what to buy, to deciding how and where to buy it.

We begin by clarifying the scope and generalizability of the research presented in this part of the report. Our research for this part included reviewing literature and case histories on corporations in the private sector. Although we include DoD examples, we emphasize private-sector experience with sourcing, an emphasis arising from the similarity of the sourcing decision faced by DON and that already confronted by many private-sector organizations. *Private industry today is rethinking where to draw the boundary lines of the organization. In the past, those lines were sharp. Now they are becoming blurred.* More and more, outside organizations have a partial presence inside the firm. It is not merely that firms are outsourcing more than they have in the past 50 years. It is that outsourcing has changed in a way that gives a greater role in complex firm activities to those who do not directly work for the firm. *We believe it is important for DON to understand this change.*

The origin of the change is the experience of U.S. automobile firms who noticed that Japanese auto companies had a very different industrial structure from their U.S. counterparts. Japanese firms coordinated fewer suppliers, but each supplier tended to have a long-term relationship wherein price of supply was only one of many factors. Instead of buying—"sourcing"—from a larger number of firms with an arm's-length relationship to the buyer, Japanese firms considered quality, reliability, and technological adaptation of suppliers[1]—a very different industrial model from the American experience.[2]

---

[1]See James C. Abegglen and George Stalk, Jr., *Kaisha: The Japanese Corporation: How Marketing, Money, and Manpower Strategy, Not Management Style, Make the Japanese World Pace-Setters,* New York: Basic Books, 1988.

[2]The Japanese industrial model has been compared with the U.S. defense sector's use of prime and subcontractors. While there are some noteworthy similarities, there are basic differences. Chiefly, defense subcontracting in the United States tends to be at arm's length, through a complex, intermediating layer of laws and regulations absent in Japan. In addition, most U.S. defense subcon-

What we can generalize from our research is that the private sector has transformed relationships to much more flexible ones; is doing more outsourcing, especially of complex work; and formulates strategically what should be sourced inside the organization and what should be sourced from outside of it. The cases and related discussion presented in the following sections illustrate what is meant by *strategic,* by *flexibility,* and by *complexity.* The basic point is that new management practices allow for a more flexible and complex set of relationships. *Again, it is this shift toward new, more complex arrangements that is generalizable from the private sector to DON.*

This formulation of the sourcing problem has shifted sourcing from being tactical to being strategic. That is, sourcing has become basic to the formulations of strategic corporate plans. How and where to source is now as fundamental as the question of what the capital budget should be, what the diversification level of the business should be, or what accounting system to use.

It is in this sense that the scope and generalizability of the research in this part of the report should be understood. We do *not assert that any particular practice that worked in industry should automatically be applied inside of DON.* Rather, the overall transformation of sourcing in the ways described above is applicable to DON. The exact application will depend on many different factors, but we believe the underlying trends are worth considerable thought from Naval managers.

Conceptions of "smart buying" or "strategic sourcing"—we use "strategic sourcing"—once centered on the trade-off between cost and quality. The field of purchasing was dominated by a debate over the conditions of when a higher-cost supplier should be selected over the lowest-cost supplier. Many factors go into such a decision, and the field focused on determining exactly what those factors were. But the new conception of smart buying goes beyond the cost/quality trade-off to ask a more basic question: Should the organization be doing this or that activity at all from the inside, or should an activity be turned over *entirely* to outsiders?[3] Thus, the term "outsourcing" is used here less to imply acquiring things from outside the organization than to resolving a basic strategic problem facing the firm: *Should the firm be in the businesses it is in, or should it focus on what it does best?* This question is taken to include not just current competitive needs, but future ones as well.

---

tracting is organized around a one-shot program, e.g., the B-1 bomber, and has little long-term joint development associated with it.

[3]There are many good surveys of the new corporate logic. For one, see Douglas A. Olesen, "The Future of Industrial Technology," *Industry Week,* Vol. 242, No. 24, December 20, 1993, p. 50.

Focus always comes up in outsourcing, because the fact that an organization can do some activity more cheaply on the inside is not necessarily a reason that it should insource. Firms are viewed in modern terms as a collection of competencies that are managed for competitive advantage. By focusing on core competencies, particularly those that are unique, i.e., have no or few competitors, firms improve their position.[4] Sourcing is embedded in a larger set of corporate decisions. Cost and quality trade-offs alone cannot be analyzed without introducing a larger theory of the firm that asks questions about what activities should be undertaken in the first place.

A number of aspects or characteristics of this shift in conceptualization illustrate what it means to do strategic sourcing in the 1990s: revitalizing the organization, increasing performance through competition, shaping and elevating the supplier relationship, drawing flexible boundaries for the organization, and managing political risks. We discuss these aspects in the following sections of this chapter.

REVITALIZING THE ORGANIZATION

It has often been said that cost savings from outsourcing are secondary to the contribution from outsourcing to revitalizing the organization. For example, outsourcing can free up scarce management attention, which can be directed to the core areas of business. It can also be used as a way to constructively dismantle certain parts of the firm.

One senior Kodak vice president described that company's purpose in outsourcing information (processing) technology (IT) and services by saying, "We wanted people who could solve problems in five days rather than five months."[5] The underlying idea was that corporate cultures can be impediments to change and can be difficult to modify through conventional management methods.

Corporate cultures seem to defy cost-accounting approaches to changing them, e.g., changes in the reward structure for purchasing agents.[6] In the mid-1980s, Kodak purchasing agents were rewarded for least cost, subject to quality con-

---

[4]C. K. Prahalad and Gary Hamel, "The Core Competence of the Corporation," *Harvard Business Review*, May–June 1990, pp. 79–91.

[5]Statement of Katherine Hudson, Vice President for Corporate Information Systems at Kodak, Harvard Business School Video Tape, Cambridge, Mass., shot in July 1991. Kodak's experience with outsourcing its IT needs is discussed in greater detail in Chapter Seven.

[6]This analysis is based on one of the author's, Paul Bracken's, management-consulting projects for a Kodak supplier, Conductron, Inc., Rochester, N.Y., in 1985.

straints. Bonuses were distributed based on the agents' ability to perform optimally to reduce costs. As a result, Kodak purchasing agents did everything in their power to drive down costs. For example, payments were withheld from circuit-board suppliers until their products were actually placed in Kodak products, even if those products sat in inventory for several months. Such practices resulted in very few of those suppliers being able to obtain bank credit for modernization, because their business prospects were not attractive, and many suppliers were driven out of business. And the technological level of suppliers that did survive often did not advance because they could not afford to purchase new equipment.

In this case, Kodak could do little or nothing to improve its relationships with its suppliers merely by altering the incentive plan for purchasing agents. Calls to think "long term" were of no avail because purchasing officers, being too removed from the strategy process, had no conception of long-term Kodak thinking. So used were suppliers to this behavior that, if long-term contracts were given out, their behavior could not have been expected to be much better. They doubtless would have anticipated a change back to the old ways and would have maximized short-term benefits.

The notion of "constructive dismantling" of such a corporate organization—eliminating it altogether or changing it by introducing new actors and practices into it from outsourcing (see Kodak example in Chapter Seven)—is key to understanding the growing use of outsourcing in U.S. industry today.

INCREASING PERFORMANCE THROUGH INCREASING COMPETITION

Having argued that outsourcing can have (positive) revitalizing effects does not mean that there are not savings. Outsourcing can reduce costs by increasing competition, both within the organization and between outside and inside suppliers.

General Motors has long used the practice of announcing that percentage cuts in capacity will take effect in one or two years' time. This practice purposefully places competing plants in head-to-head competition. The competing plants have a known period of time in which to increase quality, lower costs, innovate, and improve their performance.

The most obvious use of competition is to pit internal and external suppliers against one another. Most of the studies done on outsourcing in the public sector dwell on this issue.[7] Internal suppliers have a tendency to have a tacit long-term contract, which, from an economic perspective, lowers their efficiency because it encourages satisfycing, or collusive, behavior.[8] In many cases, they have a de facto tenure that gives them insider information that can be used opportunistically. For example, the economics of organizations developed by Oliver Williamson and others would emphasize that long-term relationships give more intimate knowledge of activities that can increase efficiency; at the same time, they give knowledge of satisfycing levels. Once these levels are met, performance levels no longer increase. Moreover, opportunism may direct efforts into just meeting these levels in the short term, with little attention to longer-term performance improvements.[9]

SHAPING SUPPLIERS

One very important issue is outsourcing's role in shaping the market of suppliers. Although firms do not have the power to alter or manipulate the market completely, the default alternative, that they have no power to shape it, is also false. Typically, the larger the buyer is, the more influence it has. The real problems in this area have arisen from a failure to think through how the market *should be* shaped. That is, many firms lack an adequate understanding of where they are going over the long term; as a result, they cannot address intelligently what they want their suppliers to look like.

Many particular examples exist of how suppliers have been shaped. Until the past few years, however, not much competitive strategy has been applied to the matter. That is, the organization applies differentiated instead of generic strategies in setting its own direction, but applies them much less to choosing its suppliers. Instead, for certain goods or services—e.g., janitorial services or bulk commodity purchases—least-cost suppliers are desired. For others, such as important electronic components, greater emphasis might be placed on technological advancement or reliability of supply.

---

[7]See Donald F. Kettl, *Sharing Power: Public Governance and Private Markets,* Washington, D.C.: The Brookings Institution, 1993.

[8]*Satisfycing* is a made-up word combining two words—*satisfying* and *sufficing.* It means just enough to satisfy or meet a demand or need; or, less euphemistically, just enough to get by.

[9]Oliver E. Williamson, *The Economic Institutions of Capitalism: Firms, Markets, Relational Contracting,* New York: Free Press, 1985.

ELEVATING SUPPLIER RELATIONSHIPS TO A HIGHER ROLE AND STATUS

Systematic attempts to address this issue have been relatively rare[10] but are increasing. There is growing recognition that supply decisions are too important to be left to purchasing agents. Instead, such decisions are being elevated to a higher role and status to form an important part of overall corporate strategy, joining finance, marketing, and operations as central to the management function. *It appears that strategic sourcing is likely to become as important in the 1990s as quality management became in the 1980s.*

It is not hard to see why this situation is likely to be so. Quality management became an important trend in the 1980s, because Japanese firms were gaining market share in the United States because they, in turn, had better-quality products. There was no way for U.S. firms to respond to this challenge by varying price or by using better advertising. These approaches were attempted, especially in the automobile market, but they were not successful in offsetting Japanese gains. Quite simply, consumers were not swayed by advertising claims not supported by experience. Ultimately, U.S. firms improved quality, in part by raising quality-management to a higher status in the firm. The message from the work of W. Edwards Deming and others was that quality-control techniques were important, but even more important was the elevation of the quality manager to a senior place in the organization.

Trends in the business environment, e.g., its increasing competitiveness, point to a similar development in the importance of strategic sourcing. There has been a great increase over the past two years in cases and articles describing how strategy can be used in the sourcing area.[11] What accounts for this increase is the recognition that some of the greatest areas of underutilized resources in the firm lie with the firm itself: in the way it does business. By shifting outside the organization activities that others could do better, significant gains in effectiveness and focus are possible.

---

[10]There are some notable exceptions. With the long planning periods necessary in the energy field, e.g., between companies selling drill technology and drill-technology users, supplier relationships have been carefully worked out. See Gregory Bruce and Richard Shermer, "Strategic Partnerships, Alliances Used to Find Ways to Cut Costs," *Oil and Gas Journal,* November 8, 1993, p. 71.

[11]See, for an illustration, Ravi Venkatesan, "Strategic Sourcing: To Make or Not to Make," *Harvard Business Review*, November–December 1992, pp. 98–107.

CREATING FLEXIBLE ORGANIZATIONAL BOUNDARIES

Sourcing decisions can also be used to open up organizational boundaries, turning them into flexible gateways rather than impermeable walls.[12] What was once a fixed feature of the corporation—its organizational structure—has been turned into a variable, subject to management and policy experimentation. Companies now can concentrate R&D efforts on their core areas, leaving technology holes for others to fill in through outsourcing. Over the next several years, we expect industry will increasingly act to maximize its technology efforts through formation of strategic alliances and outsourcing.

In R&D, such maximization is especially important, because it is in this area that new information, ideas, and approaches are most important. Duracell International, the Connecticut-based battery company, has re-directed its long-life-battery research to an alliance with Swiss and Japanese companies. This alliance has been set up to be long term and to have a high level of trust among its partners. Partners are selected both on the basis of technological expertise and on the basis of compatibility and geography. The Japanese arm is expected to monitor Asian needs and to tap into the latest East Asian technology. The Swiss arm is expected to do likewise in the European market.[13]

An ability to reach out for talent beyond one's own organization also permits acting without a number of time-consuming, many-organizational-level restraints. Quite frequently, such restraints were set up for other purposes and have developed a strong legacy value. For example, Emerson Electric Company has teamed up with Battelle to form a number of research centers focusing on advanced materials.[14] Integrated teams of researchers from each organization work side by side, united under the new relationship. Previously, cooperative research agreements would have had to pass through layers of gatekeepers coordinating the work; in the new, flexible setup, the teams control many of the decisions governing their operation.

At the same time, this reaching out beyond one's own organization reduces the need for such a large human resources (HR) department, which may have been set up, at least in part, to attract and retain the best people in the field to work for the organization. But now the need to work literally "for the organization" may be an unnecessary constraint.

---

[12]Thomas A. Stewart, "Welcome to the Revolution," *Fortune*, December 13, 1993, p. 66.

[13]Based on discussions with Robert Kidder, Duracell International, during the course of the study.

[14]Stewart, "Welcome," 1993, p. 68.

MANAGING POLITICAL RISKS

Another aspect of sourcing decisions is that they permit firms to manage political risks by hedging or even shifting them. Here, *political risks* are risks that follow from the decisions of one's own government. Increasingly, political risks are a significant part of the cost of doing business. In Germany today, for example, there is strong government pressure on large companies to create jobs in the depressed eastern part of the country. The logic behind this situation is easy to see from the German government's point of view: Job growth can win electoral support and can alleviate political tensions in a regionalized country.

But from a firm's perspective, things look different. The corporation may have certain social responsibilities, but it also has competitive demands. Inefficient job creation has to be paid for, and it limits a firm's ability to deal with other competitors not similarly constrained. In Europe, many traditional markets, such as automobiles, are being invaded by U.S. and Japanese companies, and the view of German managers is that these political costs are severely hampering their ability to compete.

Political risks are not restricted to Germany. In the United States, proposals are afoot to shift many health-care costs from the government to employers. Proposals for pension reform, occupational safety, equal opportunity, product safety, etc., abound. Corporations can respond to such cost-shifting in a number of different ways. For example, they can, in turn, shift part of their workforce to a temporary or contingent kind, thereby decreasing the size of their workforce. This approach limits exposure to new politically mandated costs. They can also outsource outside of the United States, where either U.S., state, and/or local legislation does not apply or, if it does, surveillance costs to the government are higher. In this way, firms are less at the mercy of arbitrary or discretionary interpretations of regulations. Both tacks have been pursued by U.S. firms.

U.S. firms' shifting of jobs to Mexico (as an example) should be understood not only as a way to lower wages but as a way to avoid regulations. Regulations are much less strongly enforced, and workers on the payroll do not receive the expensive benefits that U.S. workers do. Furthermore, U.S. firms can experiment with new, more flexible work processes without either union opposition or the threat of paying large termination benefits to workers who do not fit into the new setup.

The trend toward a contingent workforce in the United States—a situation in which the workforce itself is being outsourced—is a result in part of political risks. It has profound social and management implications: The idea is to establish a legally clear, arm's-length relationship between company and worker to lower the responsibility of the former to the latter. A temporary employee at IBM does not qualify for IBM benefits. Moreover, what experience shows is that new workers in this situation have lowered expectations about the paternalistic responsibilities of the firm. They are not only willing to work with fewer benefits, but their expectations of corporate day care, firm-financed education, extended leaves, and periodic review and promotion are sharply lowered.[15] There are costs as well, of course, in that employee loyalty is reduced or eliminated.

CONCLUSION

There does not appear to be a sustainable equilibrium in the relationships described here. As economic conditions change, the balance of insourcing and outsourcing can be expected to change. So too can new concepts of what it means to do strategic sourcing. All that can be done is to stay on top of these developments, recognizing that old solutions may need to change or even be replaced as underlying conditions change.

None of this analysis is to suggest that DON should or even could pursue a similar path. That is not being proposed here. What is necessary, however, is that DON be aware of the shifts in management and HR concepts going on in the U.S. market, because it will be affected by them. On the other hand, it is possible that some of the concepts described here do have applicability, or could be adapted, to DON, and senior Naval personnel should be aware of them. The most important point for DON to consider is that a major shift is taking place in both management structures and labor markets. As far as we know, as of the beginning of 1994, that shift had not been reflected in current Naval management training.

The concepts described here must at least be considered as part of the world that defines DON restructuring in a period of lower budgets. Radically new management structures and cultures follow from the trends described here and are being developed in the private sector. New models of managing the increased use of outsourced workers have been noted.[16] But a reading of DON publications reflects almost none of these concepts and developments.[17] DON acquisition publications emphasize not the trends described here but

---

[15]These findings come from case studies and interviews conducted by the author on condition of anonymity of the firms. Other research supporting the direction of these findings can be found in a 1992 survey of corporate HR trends by Towers Perrin, Inc., the New York consulting company, for its client IBM. See Towers Perrin, Inc., *Priorities for Competitive Advantage*, New York, 1992.

[16]For one of the most interesting examples, see Charles B. Handy, *The Age of Unreason*, Boston, Mass.: Harvard Business School Press, 1991.

[17]U.S. Department of the Navy, *RD&A Management Guide*, 12th ed., Washington, D.C.: U.S. GPO, NAVSO P-2457, February 1993.

compliance. Compliance with congressional laws, and with internal DoD and DON regulations, is important, but those constraints should not form the centerpiece of the Navy's approach to sourcing.

Chapter Seven

PRIVATE-SECTOR CASE STUDIES

Important new insights can be gained from looking at examples of strategic sourcing in the private sector. In this chapter, we present case studies whose function is to provide a useful heuristic device to encourage thinking about a problem in the way that private-sector decisionmakers do. Their function is *not* to generate a set of theoretical propositions about management decisions. Most of the time the underlying industrial structures are so different from one case to the next that generalizations about decisionmaking would not be defensible. Some of the public-sector cases—given in Chapter Eight, and which complement the private-sector examples here—are oriented less toward decisionmaking and more toward providing illustrations of the flexible sourcing that may be more directly applicable to the DON environment.

OUTSOURCING INFORMATION SYSTEMS AT EASTMAN KODAK

In 1989, Kodak[1] faced the decision of whether to make a capital expenditure in its varied information-systems components to modernize and consolidate its activities.[2] Initially, the question was defined, How could Kodak obtain state-of-the-art information technology (IT) at a reasonable cost? This led to consideration of different locations for a new computer facility and hiring plans for new specialists—all at a time when Kodak was locked in a major struggle with Japanese competitors in their core film and camera businesses. After a series of meetings among senior Kodak executives, this question was recast into a different one.

---

[1]See Chapters Six and Ten for additional discussion of Kodak.

[2]This case draws on Lynda M. Applegate and Ramiro Montealegre, "Eastman Kodak Co.: Managing Information Systems Through Strategic Alliances," Cambridge, Mass.: Harvard Business School, Case 9-192-030, August 1993; a related video tape (Harvard Business School Video Tape shot in July 1991, op. cit.); and background research conducted by one of the authors (Bracken for Conductron, Inc., Rochester, N.Y., 1985, op. cit.).

The new question was, Is an investment of millions of dollars to fix the in-house IT infrastructure the best use of scarce resources? That is, could the millions be better spent on digital imaging, specialty chemicals, and other core areas—to benefit Kodak's strategy for dealing with the competition? As a result of this reformulated question, an outsourcing strategy was created that involved establishing major long-term agreements with suppliers who could handle Kodak's IT needs.

Several efficiency arguments were considered by Kodak. They are described generally in Chapter Six. Of specific interest to Kodak was lowering the size of their workforce because sales per employee were $140,000 in 1989 compared with $380,000 per employee for their main rival, Fuji Film Company.[3] In addition, there were strong scale economies in certain parts of the data-processing field. One consequence of the economic shocks to the computer industry in the late 1980s was that it created within itself an oversized infrastructure, something that conceivably could be shared by outside firms such as Kodak. It is true that Kodak would have to pay a fee to outside suppliers, but the opportunity for rationalizing capacity for efficiency purposes in *two* organizations (Kodak and an outside supplier) was greater than it was for Kodak alone to retain all the work in-house.

The Kodak decision was to outsource three parts of its IT needs: telecommunications, personal computing, and centralized data processing. Contracts were signed with vendors within one year of the start of the analysis of the decision to outsource. Those contracts involved shifting Kodak personnel to the payroll of the outside suppliers, structuring the new relationship to encourage trust and cooperation, and transferring certain physical assets from the Kodak balance sheet to the outside organizations'.

A word should be said about Kodak's corporate culture. It was one in which a job at Kodak was a job for life. The firm took care of many of the employees' needs and, in effect, became an extended family. Workers would marry within the Kodak organization, and the social life of Rochester, New York, revolved around it. Founded in 1880 by George Eastman, Kodak developed over the years into one of the best examples of the paternalistic U.S. corporation. At one time, it had its own laundry, fire department, bank, and cafeteria—and most of these services were still in operation in the 1980s. In the mid-1980s, the corporate entertainment department would show feature-length films during lunch time, and returning on time to work was not considered particularly important. These practices produced a culture not unlike that in many government instal-

---

[3]C. Ansberry and C. Hymovitz, "Kodak Chief Is Trying for the Fourth Time to Trim Firm's Costs," *Wall Street Journal*, September 19, 1989.

lations, a kind of so-called bureaucratic mentality whereby legacies of past practices build up and understandings are developed about the nature of the work. In Kodak's case, the culture led to a failure to question the internal resource-allocation system of the firm. With its strong paternalistic corporate culture, the organization developed a tendency to reproduce itself, sealing off outside influences and ideas until the competitive threat from Fuji could no longer be ignored. Every resource-allocation system tends to stagnate over time, and Kodak's was no exception.

Employees were presented with three choices: find another job within Kodak; join the new strategic-alliance organizations; or terminate their employment. In the job climate of 1989 and 1990, alternative employment within Kodak was not easy to find, and certainly not in the Rochester area.

In total, about 600 workers—some of Kodak's most important employees, because of their detailed knowledge about Kodak's IT operations—were transferred from Kodak payrolls to an outside firm's. For each outside firm, an HR package was developed to provide a quality of work life comparable to that at Kodak. The average length of Kodak service for the outsourced employees was 18 years.

The significance of the Kodak IT outsourcing case is twofold. First, it marks a major example of outsourcing complex corporate operations. Most firms had few problems outsourcing cafeteria or janitorial services. But IT was different. Outsourcing IT necessitated that the outside firms become intimately involved in internal Kodak operations, and that Kodak set up new, more complicated, management structures to implement the new relationship successfully. Traditional formal contracts were not sufficient to obtain the desired results. In addition, "fluid and collaborative" relationships depending on common goals and trust were needed. (See Chapter Ten.)

Second, the Kodak case makes the point that *how outsourcing is done is just as important as the decision to do it in the first place.* Applying existing management systems to the outsourcing of complex operations is likely to prove disastrous, no matter how strong the logic for outsourcing is. Kodak undertook extensive preparation before they began to outsource. (Some of these preparations are discussed in Chapter Ten.)

It is worth noting that, despite Kodak's outsourcing, the company has not performed that well from 1989 to the present. The stock price remains depressed. Market share has still lost in important segments to Japanese competitors. And continued job cuts occur. Certainly there can be no guarantee that changes in sourcing practice alone will turn around a corporation. On the other hand, if Kodak had not outsourced its IT functions, its performance since the 1989 decision might not have been as good as it has been.

We describe the Kodak case not because it shows a beneficial effect on profitability but because it points up the need for strategic focus in the organization. *Strategic focus* defines what is and is not important. What is not key to organizational success is considered a diversion from what is. Thus, as the words imply, *strategic sourcing* transcends outsourcing and insourcing. It must cover insourcing not as any kind of "balance" to outsourcing, such as mandates that say "60 percent of depot work will be insourced." Rather, it must establish how sourcing affects the (in this case, IT) organization's value-added in what it does. The next case emphasizes just this point.

SOURCING AT CUMMINS ENGINE COMPANY

Between 1987 and 1990, the Cummins Engine Company[4] in Columbus, Indiana, faced an extremely competitive international environment. As with Kodak, it needed to become more competitive. As with Kodak, Cummins was not content to optimize its existing operations. Cummins met the need by first questioning the substantial effort it had expended during the 1980s on introducing just-in-time inventory systems and computer-aided-design machine tools, and on implementing statistical process-control techniques.[5] Stepping back from these improvements, management asked a more basic question: What effect were the areas being perfected having overall on the value added to the products of the organization? The answer was: much less than was warranted by the attendant investment, risk, and opportunity costs.

Cummins proceeded to shift corporate resources—capital and management attention—away from noncore areas, such as commodity parts (e.g., pistons), to core areas and probable future core areas. For diesel engines, the core and future core areas were judged to be ceramic engines, electronic systems control for engines, and international expansion. This shift generated a number of concerns. Chief among those concerns was the standard response both to proposals for increasing outsourcing and to moves to a strategic formulation of the problem: the assertion that outsourcing will lead to a "hollowing out" of the corporation. To help resolve this issue, Cummins undertook a study of six manufacturers in addition to itself, including John Deere, JI Case, and Navistar International.

The study had two major findings.[6] First, every firm was doing more outsourcing to remain competitive. Second, there was a pattern of initial skepticism toward outsourcing from different parts of the firm. While the different firms had

---

[4]See Chapter Ten for additional discussion of Cummins Engine.

[5]Venkatesan, "Strategic Sourcing," 1992, p. 98.

[6]Ibid.

widely varying approaches, all were compelled to outsource to meet increased competition. Many of the fears over conceptualizing the decision as a strategic one involving an analysis of what to build in-house and what to acquire from the outside had no basis in fact. For example, the term "hollowing out" was widely used to oppose everything from the outside, suggesting that the firm would become little more than a financial holding company, as have other firms. Nike, for example, does not manufacture a single shoe. But such examples are extremely rare and apply only in branded consumer goods, for which the name itself is the value of the product. In the diesel-engine business, fear of hollowing out was used to argue against something that was not being proposed. There is a strong parallel here to reactions from different parts of the defense field to similar proposals.

A more basic fear found in the Cummins example was the strategic aspect of distinguishing between core and noncore products and activities. It was easy to argue for outsourcing to save money. It was harder to decide what was of such core importance that it should *not* be outsourced. *Questioning in this area exposed the fact that there was no overall strategic perspective for an important part of corporate activities.* The actual decisions on what to outsource were made in the absence of such a framework: to utilize machine tools that were idle (Why have these tools on the books in the first place?); to shed problem parts (But didn't these offer the greatest opportunity to learn new production techniques?); and to minimize labor problems with unions (But did this confront the issue of increasing competition?).

In short, Cummins discovered that its strategy was not merely suboptimal, it was based on emotion and myths—a distorted picture that was reinforced by accounting systems. Most of the six financial accounting systems that were analyzed allocated overhead evenly across all parts, without indicating what parts added value to the overall product. One lesson that Cummins drew from its study was that some very simple decision aids had to be developed that would lead to greater strategic sense in their sourcing decisions. One such aid, for example, might be in the form of a chart that breaks the firm's activities down into subactivities and totals for each how much it would cost in money, time, personnel, and suppliers to insource—as opposed to what it would cost to outsource. (Such a chart might be useful to DON.)

We believe a major lesson from the Cummins case is that there is a pattern of opposition to strategic conceptualizations—and that such a pattern should be anticipated and managed, but should not be accepted. Were DON to adapt some of the thinking set out here, it would immediately be accused of "hollowing out the Navy"—especially in the RD&T area. There are also likely to be specious economic and operations-research studies that purport to show how an organization with slack capacity can do things in-house cheaply because the marginal costs of doing so are lower. Probably little or no attention will be given to opportunity costs, a shift to a high-fixed-cost organization that follows from not paying attention to such costs, or to consider management attention itself a scarce resource. In the Cummins context, there is little doubt that labor relations are important. But the strategic issue is not the importance of labor but whether labor is *more* important than developing new skills in ceramic engines.

Even if one decides against making this trade-off, the problem Cummins discovered was that its accounting and management systems did not properly formulate the problem in terms of new skills. Instead, the manager in charge of labor relations was asked his advice, and he said that he favored insourcing "to preserve jobs and maintain cordial relations with the union."[7] And manufacturing managers natually had a "strong incentive to insource production. After all, more parts means more responsibility, more authority, and bigger salaries."[8] In both cases there is clearly suboptimization, reflected in the structural division of the firm that gave these managers a representation that was inappropriate for taking a more strategic view of things.

---

[7] Ibid., p. 100.
[8] Ibid.

Chapter Eight

PUBLIC-SECTOR CASE STUDIES

In this chapter, we offer a set of examples illustrating strategic sourcing in the public sector; we also give one private-sector example that we believe is more appropriate to this discussion than to that in Chapter Seven. The examples show how flexible sourcing policies and other aspects of organizational culture differing from those typically encountered in DoD can lead to successful accomplishment of program goals. Although not a public-sector example, the airline industry is included in this chapter because of its similarity to one of the public-sector examples and because DoD and this industry buy aircraft from almost the same set of suppliers. In considering these public-sector examples, it is worth keeping in mind that these realizations of strategic sourcing have come into existence despite disincentives pointed out by many (e.g., in a September 1993 Defense Science Board [DSB] study of defense manufacturing[1]) as inhibiting DoD realization of strategic sourcing's potential. Such disincentives include personnel rotation and replacement policies and limited tolerance of, and payoff for, risk-taking.

ADVANCED RESEARCH PROJECTS AGENCY

The Advanced Research Projects Agency (ARPA) was established in 1958 to sponsor research on, and development of, promising technologies whose use in developed systems was either too uncertain or too far in the future to attract sufficient funding from the military departments. Types of projects funded have ranged from ballistic-missile defense and nuclear-test detection in the early years to armor and anti-armor technologies and light satellites more recently. They have covered the R&D spectrum: from research expanding the technology base in such areas as materials, electronics, and computing, to engineering development of systems such as follow-on stealth aircraft and sonar

---

[1]Defense Science Board, *Report of the Defense Science Board Summer Study Task Force on Defense Manufacturing Enterprise Strategy*, Washington, D.C.: Office of the Under Secretary of Defense for Acquisition, September 1993.

arrays towed by submarines. But, regardless of the phase through which the work had been taken, successes tended to occur when the work had a clear sense of mission or direction.

ARPA successes have come in several forms. Results of successful engineering development projects have been transferred directly to constructing and operating agencies, including the Arecibo radio telescope in Puerto Rico. Also, technologies pursued with ARPA funding have found application in systems developed by the military services, including the Air Force's FPS-85 phased-array radar and the Navy's SURTASS, a long, towed, acoustic array for anti-submarine warfare. In other instances, ARPA-developed knowledge or capabilities have been partially or indirectly applied in DoD and civilian programs.

ARPA has achieved its successes without a large in-house laboratory infrastructure—in fact, without any such infrastructure.[2] It has demonstrated that an R&D organization can have substantial influence if it acts simply as a catalyst, accelerating the development of concepts invented or work started elsewhere, and outsourcing the entire effort, except the overall management and administration.

A variety of institutional factors have contributed to ARPA's ability to pursue its distinctive approach to R&D. These factors include the following:

- An ability to attract very good people from universities, industry, the DoD labs, and elsewhere, with an appropriate civilian/military staff mix. Undoubtedly, this ability is the result of other factors on this list.

- Extensive autonomy for managers, together with short command chains and a minimal bureaucracy.

- Strong top-level support for program managers, and strong commitments of managers to programs.

- Reliance on the competence of program managers, instead of a large in-house infrastructure, to ensure involvement across all phases of the R&D spectrum.

- Rotation of knowledgeable people as programs change, coupled with a practice of limiting staff size.

- Recognition of the need for program output. (A significant amount of the RD&T effort of the DON laboratories finds no eventual application in operating systems.)

---

[2]We recognize that ARPA has often made extensive use of government laboratories to carry out its programs. And had such infrastructure elements not existed, it would have had to turn elsewhere. The point, however, is that what ARPA used, be it public or private, was not part of ARPA.

- Together with the application potential, full awareness of ongoing evolution in science and technology across the board. ARPA has exhibited a willingness to seize any source and exploit commercial interests when they existed.

- Willingness to allow contractors to choose their own approaches for arriving at a product.

It would be difficult, and not necessarily desirable, for DON to try to copy ARPA in all these respects. There would be inefficiencies in converting the current DON research-center structure to a more ARPA-like arrangement. The talents required may not be found among the current employees, who, as civil servants, would be difficult to replace. The transition could thus be long and costly.

Also, it would be desirable to retain certain comprehensive core competencies, rather than allowing them to disperse to the commercial sector. Moreover, the laboratories have an institutional memory that ARPA (and other parts of DON) does not.

Although the ARPA example should not be applied *in toto* to R&D management within DON, the Navy may still have a good deal to learn from it. Aspects of the way ARPA has handled its people are worthy of consideration for emulation, as is ARPA's balance across the R&D spectrum, with some lines of R&D intended to shape the advanced-technology base and others intended to explore applications through system development. Finally, there may be something to be gained in taking a cue from ARPA's successes by fostering, wherever possible, a clear mission orientation.

COMPARTMENTED PROGRAMS

The management of compartmented programs (also called "black programs" or "special projects") is more directly applicable to DON. The Navy itself, of course, has long managed programs of this type.

Three characteristics of the management of compartmented programs are particularly relevant.[3] First, the partnership between the sponsor (buyer) and the performer (contractor) is intimate and dominated by trust and confidence. In general, the sponsor sets goals and minimum essential performance measures

---

[3]The details of most of these programs are still highly classified; therefore, we cannot be specific here about how these characteristics have contributed to the success of specific cases. However, we assure the reader that the experience of the authors as both contractors and government officials in compartmented programs has been broad, spanning projects for DoD and the intelligence community. See Appendix B for brief biographies of the authors.

and leaves it to the contractor to establish the means to achieve the goals. That is, the program per se is not usually strongly, or over-, specified.

Second, the focus is on system development. Typically, little research or exploratory development effort is introduced explicitly. Instead, fairly mature technologies are generally brought in. The program successes tend to reflect an exquisite system integration of these kinds of technological elements, rather than a gamble that some dramatic breakthrough in Category 6.1 or 6.2 results will eventuate.

Third, compartmented programs are usually managed by a very small, competent, dedicated service or agency staff. Very infrequently do the service laboratories have a formal role. (Some program sponsors do not even have an in-house lab.) Instead, on occasion, individuals from the labs are briefed, their advice is solicited, and they are debriefed.

To benefit from the example set by compartmented programs, DON need not adopt as Spartan a management structure. An R&D management team could be created to act as a technology facilitator for certain initiatives. The team might include research laboratory or center individuals and have the objective of directing the attention of technologists to warfare innovations. The team's efforts would span the research, development, and user-introduction phases. Artificial barriers (6.1, 6.2, etc.) would be de-emphasized. (Ongoing acquisition reforms would have to include measures to permit such an arrangement, because current regulations could present legal constraints.)

The compartmented-program example is an apt one for several reasons. Clearly, the attributes of strategic sourcing described in Chapter Six are reflected in such characteristics of compartmented-program management as intimate buyer-supplier relations and focus on core organizational activities. Further, the success of such programs raises questions about the need for an extensive DoD technology infrastructure. Finally, most compartmented programs are tuned to low procurement quantities and thus offer lessons in the kind of "lean production" that is anticipated to be the mode for much future DoD procurement.

AIRLINE INDUSTRY

Although the airline industry is not in the public sector, we discuss it here for two reasons. First, its sourcing strategies are similar to those of compartmented programs; second, both the civilian airlines and DoD act as buyers from the same set of aircraft-industry suppliers.

As with compartmented programs, the airline industry has no in-house laboratories and supports hardly any research or exploratory development (an exception being noise-reduction R&D). Its focus is on the development and purchase of engineering services, the objective being that the supplier will provide a virtually complete, ready-to-operate product. The airline sets the requirements, and industry develops a product to meet those requirements. Because the buyer does not micro-manage the supplier's activities, the supplier can maintain a smaller procurement organization.[4] Nonetheless, mutual interests and practical considerations have resulted in a great deal of interaction between buyer and supplier, so that the two function almost as partners.

The potential of this model for DoD aircraft procurement is particularly intriguing because, as mentioned above, the main civil-sector aircraft suppliers also supply the military.[5] About 20 percent of Boeing's sales and 60 percent of McDonnell Douglas's (MD's) sales go to the U.S. government.

Of course, financial arrangements within the military-aircraft market differ from those within the civilian sector. In the military market, the government shoulders the bulk of the R&D risk. In contrast, the civil fleet operator often has a share in funding major new production programs, and most of the risk and financial obligations are borne by the designer and manufacturer of the aircraft. The civilian-sector buyer may even seek purchase-price warranties. Furthermore, the whole relationship of the buyer to the production program is different. DoD commits to buy a number of units sufficient to enable the program to be undertaken; in some cases, that number is all the units that will be produced. The civilian airline buys on the margin as many units as it needs and may postpone purchases or substitute upgrades as the market and its own financial situation warrant.

In other respects, however, there are similarities between the two sectors. Their market sizes are on the same order of magnitude. Military sales have varied between roughly one-third as much and half again as much as civilian sales (usually closer to the former figure) over the past 15 years. The complexity of the markets and products is also similar. Product diversity is about equivalent, and, we assert, superficial impressions to the contrary, so, generally, is the degree of engineering sophistication; in fact, if the U.S. civil sector develops a hypersonic or longer-range, high-speed civilian transport, its engineering

---

[4]It is of interest that the aircraft manufacturers themselves pursue a flexible sourcing strategy. Jet engines, for example, are bought from an outside supplier. Most aircraft producers have no major stand-alone jet-engine laboratory infrastructure. Instead, they rely on a small, expert engineering staff to serve as "smart buyers" for these complex products.

[5]Production facilities serving the civilian and military sectors are segregated.

sophistication could be judged to surpass that of the military side in some of the associated areas, such as power plants for hypersonic speeds.

Both sectors also face an uncertain future. On the military side, this uncertainty needs no elaboration. But even in the civil sector, the U.S. aerospace industry faces increasing challenges—from abroad. Whereas the United States retains both market and technology leadership, its position has eroded. In 1970, U.S. firms accounted for almost 80 percent of the global aerospace market (excluding the communist nations); by 1993, that share had shrunk to 60 percent. Thus, the utility of retaining flexible sourcing in the face of market erosion may have some implications for the downsizing military sector.

CONCLUDING OBSERVATIONS

The foregoing examples support a major theme that is worth reiterating: Technology alone and in itself is generally not the key to the desired answer; rather, having the skill and imagination needed to apply that technology to an *integrated* system for a specific *mission* gives rise to the real accomplishment. A perfect example in its time was the U-2 aircraft: Essentially no really advanced technology was present in that system. Yet the integrated system was a breakthrough of immense military and strategic value.

The traditional categorization of defense R&D (i.e., 6.1, 6.2, etc.) does not serve well to emphasize the value added through integration and mission application. Because of that categorization, and because defense RD&T actors are usually assigned to discrete activities within those categories, there is little freedom or opportunity for life-cycle involvement in providing value-added and affordability to a user. "Smart buyers" would be specifically empowered to maximize the advantages inherent in such attributes.

EFFICIENCY GAINS FROM USE OF
CIVIL-SECTOR PRODUCTION

Flexible, strategic sourcing requires that the military buyer in the evolving procurement environment have a comprehensive knowledge of the products the commercial sector has to offer. As mentioned in Chapter Four, we judge that, over the next few decades, there will be relatively little change in most basic platforms (submarines, other ships, aircraft), but many supporting warfare technologies (computers, communication systems, etc.) will continue to evolve markedly and will continue to be driven by commercial markets. If so, incentives for reliance on civil-sector products will only grow. In this chapter, we show that sourcing strategies relying on civil-sector production can lead to gains in efficiency.

Ours is hardly the first call for more military reliance on products from the commercial sector. For example, according to a 1986 Defense Science Board Summer Study,

> commercially available computers/radios and displays are as durable in harsh environments, several times cheaper, five times easier to acquire, and more reliable than counterparts used in the military.[1]

While there are occasional military needs for specialized products with little immediate commercial relevance (e.g., very-high-pulse-power devices useful for partially simulating classes of nuclear-weapons effects), a rapidly increasing commonality between the two sectors is evident. For example, a vast class of information-related devices with dual uses has come into being.

A principal reason why previous calls for greater use of commercial-sector products have not been answered lies in the presence of a set of generic imped-

---

[1]Defense Science Board, *Defense Science Board 1986 Summer Study: Use of Commercial Components in Military Equipment*, Washington, D.C.: Office of the Under Secretary of Defense for Acquisition (OUSD[A]), January 1987. A subsequent study was done in 1989: DSB, *Report of the Defense Science Board on Use of Commercial Components in Military Equipment*, Washington, D.C.: OUSD(A), June 1989.

iments characteristic of military procurement: excessive technical data requirements; obsolete, costly specification procedures; a preoccupation with audits in place of process development and optimization; etc. The value of easing some of these impediments should be clear from the following examples of the role that commercial products could play in flexible, strategic sourcing.

MICROELECTRONICS R&D

The ability to design and manufacture advanced integrated circuits (ICs) from semiconducting materials is the foundation of the information age. U.S. defense policy rests on the qualitative superiority of U.S. weapon systems, which, in turn, rests on the superiority of the electronics incorporated in those systems. The health of the microelectronics industry in the United States has been of concern to industry observers and policymakers for the past several years. The prevailing view is that increasing competition from abroad has led to a decline of worldwide U.S. market share.

The U.S. government, and especially DoD, has played a significant role in funding microelectronics R&D and thus in developing the semiconductor industry. However, most of the money spent on microelectronics R&D in the past decade has been from commercial semiconductor firms on R&D related to commercial markets. The R&D funding provided by the government during that period was either for basic research or for products and processes of interest to the government, e.g., to improve the performance and availability of advanced military microelectronics. Little coordination appears to have taken place between government-funded R&D and development under way in commercial markets.

To determine whether government R&D funding resulted in microelectronic devices superior for defense purposes to those developed in the private sector, RAND undertook a study of the development and performance of four product groups: microprocessors, digital signal processors (DSPs), static random-access memories (SRAMs), and programmable read-only memories (ROMs).[2]

The study's two principal research questions and the answers to each can be paraphrased as follows:

- Can commercial markets be expected to lead or equal military markets in the introduction of technologically advanced products?

---

[2]Anna Slomovic, *An Analysis of Military and Commercial Microelectronics: Has DoD's R&D Funding Had the Desired Effect?* Santa Monica, Calif.: RAND, N-3318-RGSD, 1991.

The answer is yes in every category of components. The military has, for the most part, been behind the commercial market in development and implementation. In some components, like microprocessors, the military is definitely behind the commercial market. In DSPs, ROMs, and non-rad-hard SRAMs, differences are not large.

- Does the government's strategy of skipping product generations produce advanced ICs faster than commercial evolutionary development?

No is the answer supported by data for all four component groups. There is no indication that ICs available to the military are different from those resulting from adaptation of commercial components and technologies to military applications.

In the case of microprocessors, for example, commercial products have equaled or surpassed their military counterparts in miniaturization, performance, and price/performance ratio (see Slomovic's Figures 9, 12, and 14, respectively[3]). As a result, in microprocessor and other microelectronics R&D, very few areas of DoD leadership remain.[4] Spin-offs from military R&D to commercial use are virtually nonexistent. Instead, civil-to-military "spin-on" is becoming the paradigm, for example, in the use of commercial reduced-instruction-set computer (RISC) processors as military standards.

It thus appears that the most efficient approach for DON to take in securing further advances in microelectronics is to focus on translation of commercial advances, e.g., submitting commercial products to more extensive military packaging and testing requirements. However, close attention should be given to the need for such additional requirements, because they are what drive up the cost of products and the time it takes to bring them to market. It is partly because of such requirements that firms participating in the military microelectronics market either do not participate in the commercial market or do so with divisions separate from their military divisions. In supporting separate production facilities that could not compete in the commercial market, DoD pays a high price for its additional demands. If such demands could be relaxed to permit acquiring microelectronic products from firms and divisions having a commercial IC mass-production capability, DON could take advantage of the competitive incentives for cost reduction that exist in the commercial market.

---

[3]Ibid., pp. 58, 60, 62.

[4]Much DoD-only development effort has gone into hardening microelectronics against nuclear-weapons effects (making them rad-hard). The resulting product's capabilities far exceed those of standard commercial-world microelectronics, but the product is very expensive. Because of the nature of the Cold War threat, it is plausible that the effort was worth the money. However, in the new, post–Cold War world, the effort's usefulness is less clear.

FS-X RADAR

Commercial-sector efficiency gains can also be realized by making fuller use of commercial production *processes*. This fuller use is illustrated by the Japanese approach to weapon-system acquisition, and particularly by the development of the advanced phased-array radar for the FS-X, a fighter aircraft developed in collaboration with the United States.

The FS-X radar relies heavily on gallium arsenide (GaAs) technology, a technology that the Japanese have been developing for commercial purposes for over 20 years. It was applied first to televisions, videocassette recorders, and compact-disc equipment; then to cellular telephones; and now in automobiles, traffic-control systems, avoidance sensors, and general telecommunications. Japan supplies most of the world's GaAs materials, and the United States is a major purchaser. Some of that demand is military, e.g., for the upcoming F-22 phased-array radar, but cellular-telephone usage alone will surpass that for the F-22.

The lesson here is that *the Japanese have taken the lead in a highly sophisticated technology with important military applications, because they have integrated military R&D and production with R&D and production aimed at the larger commercial market.* Granted, the FS-X radar will not perform to the specifications of the F-22 radar, but that may be a deliberate choice. The FS-X radar is designed for producibility, high commercial content, and far lower cost than the F-22 radar, and it is considerably ahead of the F-22 radar in schedule. The Japanese philosophy seems to be to continually improve to higher performance by product and process evolution, rather than by a technologically riskier attempt at a major single jump.

The difference in approach could hardly be greater. The United States takes a performance-driven R&D approach focusing on the product, making a weak transition to manufacturing at low volume on a separate military production line; the Japanese take a cost-driven R&D approach focusing on the process, making a strong transition to manufacturing at high volume on dual-use production lines. The Japanese approach enables significant economies of scale and scope, with learning experiences available to both military and civilian products. The Japanese derive automation and flexibility incentives from large production lines, in contrast to the labor-intensive assembly and low-capacity utilization of U.S. production facilities (as illustrated by the U.S. program for monolithic microwave ICs, or MMICs).

The implications of the Japanese approach were recognized (if belatedly) in the September 1993 DSB study of defense manufacturing, which advocates "lean manufacturing"—defined by the DSB as focusing on process improvement and attention to the entire life cycle, from requirements determination, through R&D, to product support and then phase-out.[5] The DSB study also noted carefully the constraints imposed on DoD by the U.S. Civil Service personnel system, estimating that "DoD has an excess of as much as 25 percent in areas that should be affected by downsizing."[6]

If elements of the Japanese approach are adopted, there may be substantial scope for efficiency gains in DoD's acquisition of RD&T. Dual-use production-sharing with high-volume commercial applications could provide the means for improving military technology within tight budgets. But it requires increasing interaction between RD&T and industrial and acquisition management, i.e., increased interconnections between the technology and production bases. And it would also, of course, require review of military specifications and revision of acquisition regulations.

VERY-HIGH-SPEED INTEGRATED CIRCUITS

DoD's very-high-speed integrated circuits (VHSIC) initiative combines some of the points made above in discussing microelectronics R&D in general and the Japanese R&D model. This initiative was a billion-dollar effort to keep DoD on the leading edge of advanced IC technology and was undertaken in response to what DoD perceived as deficiencies in obtaining advanced ICs for military use.

At the start of the program in 1979, the Pentagon told potential VHSIC bidders that it preferred to award contracts to defense electronic-system suppliers rather than to IC producers, believing that doing so would promote use of the new chips in fielded systems. However, system houses were encouraged to team up with commercial semiconductor suppliers to ensure wide dissemination of outputs within the defense-supplier community. Nine initial contract winners were eventually narrowed down to three contractors, whose chips generally met feature size and integration density specifications by the late 1980s.

A number of commercial IC manufacturers did not wish to participate in the program because of limitations imposed by the specialization of VHSIC chips and by security restrictions. Those restrictions applied not only to the ICs themselves but also to process innovations associated with them. They were imposed by DoD because it was afraid that premature commercialization would allow the Soviets to duplicate the advances made in the program.

---

[5]Defense Science Board, 1993, p. 8.

[6]Ibid., p. 11.

Companies that did not choose to participate in the DoD program and some of those dropped in the down-selections have been actively pursuing their own very-large-scale integration (VLSI) programs, often along lines similar to VHSIC but on more relaxed schedules. Both Intel, which never participated, and Texas Instruments, which was dropped after the early 1980s, have successfully sold VLSI chips qualifying under VHSIC standards. Contractors that continued to participate have created extensive parallel commercial efforts to avoid the DoD restrictions.

These non–DoD-funded efforts testify to the commercial potential of VHSIC-grade chips and to the possibility that DoD need not have supported this major RD&T effort to begin with. They demonstrate the ability of the commercial sector to stay even with or outpace DoD-funded efforts, and the inefficiency of imposing restrictions or requirements beyond those prevailing commercially.

DEFENSE INTELLIGENCE

As pointed out in a recent OSD-sponsored RAND study,[7] even in the specialized military field of intelligence-gathering, the technological and economic value of exploiting commercial resources will grow rapidly. Gains from existing commercial systems may well compensate for any reductions in RD&T on military intelligence-gathering systems. Commercial resources are of two principal types: remote-sensing satellites and open, on-line databases. The potential of the first of these is well understood; we focus here on the second.

The magnitude of information available in commercial databases is already enormous. In 1991, DIALOG[8] alone stored 1.4 terabytes, or about 400 million pages, of information. About 5000 databases are available in the United States, and as many more exist in other countries. A great deal of information is being converted to, or is originally, digital and electronic, often supplanting all paper records. Electronic data interchange and commerce are growing explosively and globally; all types of physical infrastructures are increasingly controlled by information systems. Every country of interest has vital information available on-line, including maps and mapping imagery, technical and prototype engineering data, information on economic and development trends, and data on military structure and systems. Undoubtedly, new sources will appear, making

---

[7]The study is classified and is therefore not cited here; we offer a summary of unclassified aspects.

[8]DIALOG is an on-line, remote-access database subscription service originally offered by the Lockheed Corporation, subsequently by an independent company under this service mark, and now part of more-extensive information services being planned for public offering by the Knight-Ridder consortium.

reliance on open-source materials even greater and more productive. Clearly, use of commercial databases will be an increasingly powerful tool for military and national intelligence and threat assessments.

Chapter Ten

IMPLEMENTATION ISSUES

There have been many instances of shifts to more strategic sourcing in the private sector (a few of which are discussed in Chapters Six and Seven), and the implementation of these transitions may carry lessons for DON. In the following discussion of implementation, we focus primarily on the role management plays in implementing transitions. Both leadership and management are important, as is recognizing the distinction between the two.[1]

Leadership deals with change; *management* deals with the complexity of implementing the directions that leadership points to. Both leadership and management are needed to change an organization. Leadership without management is empty and abstract. Management without leadership is unexciting, uninspiring, and likely to pull in too many different directions at the same time. It takes leadership to see sourcing as strategic—for example, to go beyond just consulting managers in charge of production or labor relations, who will probably advise against a sourcing change. But it also takes management to carry out the new direction, to make it actually work. Conceived in these terms, one of the first questions that should be asked when attacking any strategic problem is, Which individuals on the team have leadership skills, and which ones have management skills? Again, implementation requires both.

The leadership-management distinction leads to the first of several points concerning implementation of outsourcing: *Outsourcing strategies that are not thought out from a management viewpoint will be disastrous.*

For the kinds of complex issues and operations discussed in this report, a memorandum mandating the Navy to outsource more—if that is all that is done—will produce great confusion and opportunities for mischief, as did directives for least-cost in the private sector. Evidence from nondefense, public-sector cases supports this conclusion.[2] In the private sector, there is a natural

---

[1] See J. P. Kotter, "What Leaders Really Do," *Harvard Business Review*, May–June 1990, pp. 103–111.

[2] Kettl, *Sharing Power*, 1993, pp. 1–20.

tendency for firms to conceal their disasters and to not allow them to be written up as business-school cases.

In practical terms, we hope that the above italicized observation stimulates the Navy to specify a management plan to carry out any outsourcing that it chooses to pursue,[3] and that different skills—management and leadership—are included on the implementation team. The management plan might provide for some of the elements discussed individually in the following sections.

TRANSITION ORGANIZATIONS

One method used by Kodak managers to implement outsourcing was to separate the long-term management structures for outsourcing from *transition organizations*, those structures needed to put outsourcing in place. Recall that in the Kodak case a great emphasis was put on the HR aspects of transferring employees from Kodak to supplier payrolls. The transfer was a one-shot affair, and there are reasons for not continuing the organization that carries out what may be especially controversial decisions.

Transition organizations can include inside and outside representatives who are tasked with making things happen on schedule and on budget, taking broader directives as guidance. They may be more like teams than committees, i.e., they may be given a budget to control, may be empowered to make decisions, and may be held responsible for the results.[4]

ARCHITECTURAL KNOWLEDGE TEAMS

A special kind of team is the one tasked to stay at the forefront of the outsourced activities. It is called an *architectural knowledge team* because it retains, in its members' collective memories, the detailed knowledge of the latest technologies and processes related to the outsourced activity. Such teams can be located at the home office or in the offices of suppliers. All team members work for the firm doing the outsourcing, and it is important to retain good people in these jobs.

---

[3]It might be years before a change in RD&T sourcing policy is felt or observed in the aggregate capabilities of the Navy. On the positive side, it might be reasonable to take more risk because, in the short run, it would not cripple the Navy's capability to perform its mission. On the negative side, when long-run impacts show up, it may be difficult to correct the situation. Consequently, the Navy would need to assess the impact at a low level in the organization, not just at the total Navy level. The Navy would need to explore explicitly ideas on how it could monitor the effects of sourcing changes, i.e., what to look for, how to measure it, and how to assess when a bad decision had been made.

[4]For the difference between teams and committees, see D. Quinn Mills, *Rebirth of the Corporation*, New York: John Wiley & Sons, 1992, pp. 29, 133–135.

Cummins Engine outsourced virtually all its piston manufacturing. Yet in its architectural knowledge teams it retains some of the foremost experts on piston technology.[5] A point worth emphasizing is that it is not necessary to be *in* a business that has become commoditized to keep *abreast* of developments in it. The reverse could prove a very expensive way to stay on top of new technologies.[6]

FLUID CONTRACTING

Sourcing decisions, with their emphasis on cost, quality, delivery date, etc., are subject to a great deal of legal scrutiny. One of the first things the Kodak management did was to ban lawyers from some early meetings between suppliers and the firm, because they wanted to establish a long-term relationship, not one defensible in court in the short term.

The term "fluid contracting" has been coined to describe contracts that emphasize long-term joint trust rather than litigation threats to enhance short-term performance. Both types of contract have their place, depending entirely on what the firm wants to do. Fluid contracting is a recent development, and new understandings are needed on how best to carry it out. It entails certain shifts, such as contract outlines being specified by managers, not lawyers; consultative commissions for dispute resolution; alternatives to litigation; and structuring teams that have an interest partially transcending organizational interests.

TYPING SUPPLIERS FOR PARTNERING CHARACTERISTICS

The types of suppliers a contracting organization will work with need a great deal of thought. In addition to questions about technical and financial viability, similar (or different) corporate culture, access to technology, energy, experience in outsourcing relationships, and most important, commitment to a long-term relationship, must be considered. In both the Kodak and Cummins cases, and others, matrices of desirable characteristics were drawn up in advance and a scoring system was used to rank candidates.

---

[5]There is a risk here, of course, in that in order to stay expert, it is generally necessary to "keep one's oar in."

[6]See Rick Bleil, "Increasing Competitiveness Through Better Supply Management," *Electronic Business Buyer*, Vol. 19, No. 11, 1993, p. 72.

PROBLEM-DEFINITION SEMINARS

Focusing on the difficult management aspects of outsourcing implementation, prepared, problem-definition seminars have proven useful in communicating some very complex, and potentially threatening, decisions. Subject matter can range from HR to rationales for change. Their intent is to build support for change and to clarify, for middle managers, what may be clear to senior managers but is less so the further one gets from the top of the organization.

NEGOTIATION AND BROKER ORGANIZATIONS

Interorganizational communication is problematic under the best of circumstances. On issues of outsourcing, in which jobs are at stake, it can be even more difficult. Some firms, Kodak for instance, have hired outside consulting firms to smooth the process. Broadly speaking, what these consultants do is to elicit the real utilities and interests of the involved parties—for example, to test whether a long-term relationship is truly sought.

In addition, role-playing games have proven useful to study the response to various possibilities, e.g., how much burden can be shared in an economic downturn and whether it is likely to rupture the alliance.

EXECUTIVE EDUCATION

To properly carry out an outsourcing strategy, as distinct from a one-time outsourcing contract, requires commitment from all levels in the firm. Too often there is a tendency for senior management to "wing it," leaving to lower-level managers details they have not been trained to deal with. Not knowing what is expected of them and having little or no experience base to tell them what problems to look out for, lower-level managers, even if they support the changes, may be unfamiliar with the tools and attitudes needed to make them work. For example, the point in an outsourcing initiative at which HR effects are discussed with employees is a critical one, but its importance may escape the attention of senior managers. Experience suggests that supervisors will receive many questions from workers, and it is important for them to have some form of an answer.

Executive-education programs can boost the confidence of implementers. They are not a panacea, but they can be a valuable tool to facilitate communication of managers with each other and with workers.

Chapter Eleven

CONCLUDING OBSERVATIONS FOR PART II

THE RATIONALE'S THE THING

If Part II of this report does nothing else, it should enlarge DON's thinking about rationales for outsourcing. The private sector in the United States has greatly increased its use of outsourcing. In Part II, we argue that the rationale underlying this trend is much broader than short-term cost savings; it ranges from constructive dismantling of existing organizations and corporate cultures to reducing or eliminating political risks. To define the need for outsourcing in terms of cost reduction alone is to enter a technical economic discussion of "true" overhead rates—rates that include value added by a particular line or product, as discussed in relation to Cummins Engine at the end of Chapter Seven. And we have yet to see a single example in the private sector of the technical calculation of overhead rates driving a management decision to outsource.

Yet this is the character of the debate that informs public-sector outsourcing. Technical, economic arguments involving game-theoretic calculations to allocate overhead shares certainly have their place, but in this particular field of public policy analysis, it is striking that the most important reasons for outsourcing in the private sector are little debated by the Navy.

Outsourcing comes from an understanding of why an organization sources as it does. In many examples familiar to us (e.g., Kodak's failure to question resource-allocation practices), the sourcing reason is based on myth, emotion, and legacy rather than on rational thinking and analysis.

THE DON AND STRATEGIC SOURCING

DON would have to ask itself what is most important, or core, to it, and what is not. Activities that are not core are candidates for outsourcing. Activities that *are* core to it can be done either inside or outside, depending on many variables. In a field such as R&D, many examples suggest that keeping too much work in-house erodes contact with new ideas and technologies. The in-house organization develops a "not invented here" syndrome (i.e., if it's not our idea, it's probably not worth considering), or begins to add superfluous support activities that divert attention from its original purpose.

Strategic sourcing—an approach whereby an organization first determines what is most important, then makes associated decisions about where to source—can be a therapeutic management exercise with unexpected positive results in revitalizing the organization and achieving focus. Its application to DON will probably be imperfect and incomplete. But in attempting to apply it, department decisionmakers, by focusing on areas key to what the Navy does and how it does it, have the potential for yielding great benefits to the Navy.

PART III

COMBINING PARTS I AND II AND DRAWING INFERENCES

SUITABILITY OF RD&T LINES FOR OUTSOURCING AND POSSIBLE IMPLICATIONS OF THE FRAMEWORK FOR THE NAVAL RD&T INFRASTRUCTURE

This chapter combines the ideas given in Parts I and II to offer a speculative sketch of how DON might think about which of its RD&T lines are more suitable for outsourcing and might infer which institutions and facilities are better candidates for realignment or retention.

SUITABILITY FOR OUTSOURCING

The support priorities framed in Part I and the sourcing strategies discussed in Part II should not be considered in isolation. Anticipated relative value to DON and anticipated extent of future breadth of demand clearly impinge on decisions to be made about where to procure products and services in different lines of RD&T. In principle, even high-relative-value, Navy-unique lines of RD&T could be outsourced. However, in practice, a better case can be made for retaining RD&T capability in-house for those lines whose future RD&T is anticipated to have a high relative value than can be made for those lines whose value is judged to be less; at the same time, flexible sourcing options would be more varied when the anticipated future demand (and therefore supply) is broader.

Thus, regions of Figure 4.2 that are suitable for flexible sourcing might resemble those illustrated in Figure 12.1, where the lighter the shading is, the greater is the suitability for flexible sourcing.[1]

However, two provisos should be attached to this reinterpretation of Figure 4.2:

- Clearly, the gradient is not as strong, nor necessarily in precisely the same direction, as it would be for funding support.[2] That is, it is more apparent

---

[1] Except that the shading is different, Figure 12.1 is the same as Figure 4.2.

[2] Although we do not use the word *gradient* in Chapter Four, the discussion in footnote 6 of that chapter is about gradients and gives their general directions as "along lines that slope downward to the right" in Figure 4.2.

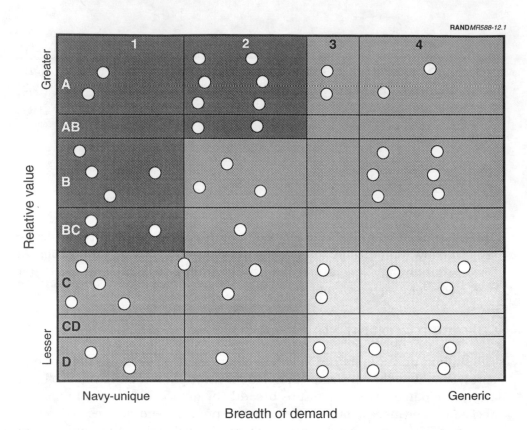

RAND*MR588-12.1*

NOTE: The lighter the shading is, the greater the suitability for flexible sourcing.

**Figure 12.1—Inferring Suitability for Flexible Sourcing from Relative Value
and Breadth of Demand**

that RD&T categories toward the upper left need to be supported *somehow*
than that they need be supported in-house.

- When reliance on commercial-sector products is specifically sought, it
  follows from the definition of classes of breadth of demand (see Chapter
  Three) that the only suitable candidates are those on the right side of the
  diagram.[3]

Again, as emphasized in Part I, the specific definition of RD&T lines and
assignment of those lines to locations on the chart would require DON input.
Were the Navy to construct its own version of a finished chart like that shown in

---

[3]As noted in the preceding paragraph, this statement does not imply that the gradient is horizontal
along the boundary between breadth-of-demand classes 3 and 4. In other words, the higher the
relative value in class 4, the stronger (although perhaps still weak) will be the case for insourcing.

Figure 12.1, it would have a basis for guiding decisions about where flexible sourcing could substitute either partially or completely for elements of the in-house RD&T infrastructure, especially when many RD&T lines are positioned toward the lower right of the diagram and represent institutionally or physically coherent elements of the infrastructure.

POSSIBLE IMPLICATIONS FOR THE NAVAL RD&T INFRASTRUCTURE

Thus, our framework might be extended to support decisionmaking regarding which select research facilities would continue as integral parts of the Naval RD&T infrastructure. This extension could identify those facilities that, by pursuing the highest-relative-value and narrowest-demand technologies, warrant the highest support priority. It could also highlight those technologies that warrant lower priority. With the exception noted below, the activities of these lesser-contributing facilities could either be outsourced or their support could be terminated. Either way, such facilities would be candidates for realignment or closure as part of the FY 95 or subsequent BRAC rounds,[4] or as part of less formally undertaken infrastructure changes.

A Systematic Approach

The exception is those institutions or facilities pursuing a mix of variously ranked lines of RD&T. Such institutions or facilities could either be realigned or the corresponding research shifted elsewhere. For example, to facilitate visualization of the support priorities of the RD&T lines that each Naval RD&T facility is pursuing, a matrix such as that in Table 12.1 could be constructed, with RD&T lines in support-priority order as row stubs, and current RD&T facilities—laboratories, contractors, universities, etc.—as column headings.

In the hypothetical example in Table 12.1, Facility 3 is clearly pursuing technologies having high Navy support priority, and Facility 1 is not. Facility 2 has a more mixed agenda. Facility 4, since it is pursuing technologies in breadth-of-demand class 2, i.e., also of interest to other DoD entities, would be a candidate for consolidation with another service or government agency or for assuming the lead role for DoD. Using this simple model, those facilities pursuing the highest-support-priority RD&T lines, and those that are not, are easily identified. Such an analysis may also reveal gaps in coverage of existing facilities and institutions and the need to establish new or realigned ones.

---

[4]BRAC 95 is the last round called for under current law. Nevertheless, rumors abound that additional rounds will be needed if the DoD tooth-to-tail ratio continues to fall, as is generally expected.

Table 12.1

Hypothetical Categorization of Facility Research Agendas, by Support Priority

RD&T Line	Facilities			
	Facility 1	Facility 2	Facility 3	Facility 4
Navy Sensors and Signal Processing			X	
Mine Warfare		X	X	
.				
.				
.				
Fixed-Wing Aircraft				X
.				
.				
Biotechnology		X		
Environmental Quality	X			
Fuels and Lubricants	X	X		
Civil Engineering	X			

Systematic approaches such as that represented in Table 12.1 may lead to over-estimates of the number of RD&T lines that DON can continue to support, or at least the number it can retain in-house. If DON does not manage proactively the impacts of the increased magnitude and pace of DoD downsizing, Naval RD&T products will be impinged upon in ways that are not necessarily prudent and rational. In the current near-crisis atmosphere, fairly rapid RD&T cost reductions appear inescapable in the near term, giving high priority to still-more-effective use of declining resources.

A More Draconian Approach

A more Draconian approach than that just sketched would shift nearly all elements of RD&T represented by the DON lab/center infrastructure in categories 6.1 through at least 6.3A to flexible sourcing. Exceptions would have to be explicitly defended.

And it is quite plausible that there would be defensible exceptions. For example, the Naval Research Laboratory and The Johns Hopkins University Applied Physics Laboratory are highly productive, world-class institutions. It would probably be in DON's interests to preserve their major elements.[5]   In

---

[5]Of course, there are elements of others the Navy would want to save, such as, but not limited to, parts of the four warfare centers and the Applied Research Laboratories of Pennsylvania State University and the University of Texas. Also, as suggested in Chapter Four, at the 6.1 level the preservation need not necessarily be as an activity of the Navy per se.

fact, they might serve as models for future in-house and nongovernment RD&T organizations. Such retained infrastructural elements should have at least the following characteristics:

- They have a retained corporate memory and a sense of institutional history (not always present in government organizations with high turnover in senior positions).

- They conduct core basic research, with participation in select Navy-unique lines of RD&T (e.g., SSBN security and torpedo-quieting).

- They are subjected to enhanced program review and opportunities for competition.

- They have expanded access of personnel to academic career paths.

TRANSITION ISSUES

If drastic downsizing occurs, systematic or otherwise, addressing transition issues would be of utmost importance. For activities subject to outsourcing to be truly "smart," maintaining a cadre of personnel well-versed in strategic sourcing would be critical.[6] And moving the DON infrastructure into closer relationship with the industrial and academic communities would be essential. How best to achieve these goals in the process of internal downsizing is a complex problem that has already been faced by private firms making transitions to more outsourcing as they have downsized. As noted in Part II, DON may be able to find some solutions in the lessons those firms have learned.

The approaches given above are only two of several possibilities, and approaches could be mixed. In a period of DoD budget crisis, a drastic approach may have the merits of bringing together considerations of pace, coordination, and one-time restructuring. Naturally, any approach will have costs and benefits and will draw on leadership and political capital. As discussed in Chapter Ten, political capital will be required both to secure the relaxation of regulatory and other legal constraints impinging on DON and to successfully implement changes within the department's institutional culture.

---

[6]The ARPA model, discussed in Chapter Eight, suggests that such outsourcing could be done without, for example, in-house labs, provided appropriate management structure and people were in place.

AGGREGATED FORM OF THE LONGEST LIST
OF RD&T CAPABILITIES

This appendix shows all the individual items in the longest list and their aggregation into the 53-line/capability list given in Table 2.1. The list is organized alphabetically, by entries in Table 2.1, which are shown here in bold. The longest list includes some subordinated RD&T capabilities/lines and all anticipated acquisition programs that we know of.[1]

This appendix is included in the report mainly to assist the reader in understanding three aspects of the list that will help grasp the list-making and priority-setting processes.

First, the items, programs, etc., under an entry on the list in Table 2.1 are not equivalent to the entry but only aid in getting insight into the kinds of RD&T capabilities we include in it. Thus, as examination of the list shows, a program can appear under several entries of differing ranks. For example, JAST is included under four entries—**Fixed-Wing Aircraft**, **Air-Vehicle Signature Control and Management**, **STOVL Aircraft**, and **Carrier-Unique Aspects of Fixed-Wing Aircraft**—which are ranked in different cells in Figure 4.3: BC2, A2, A2, and B1, respectively.

Second, and related to the first, we included all items under each entry in Table 2.1 to shed light on subject matter in the corresponding RD&T category. Thus, e.g., JAST's being under **Air-Vehicle Signature Control and Management** means that those parts of the JAST Program associated with this RD&T capability have the corresponding relative-value and breadth-of-demand rankings. When looking at a *program* under one of the 53 entries below from Table 2.1, the reader should not think of the program per se but of the subject-matter part of the program that is appropriate to the entry from Table 2.1. As explained in the main text, as the ranking process proceeded, we made extensive, regular use of the evolving appendix to remind ourselves of just that.

---

[1]By *subordinated*, we mean such lines as Shallow-Water ASW (which is not one of the 53 in Table 2.1), included under **Navy Sensors and Signal Processing**.

Also, together with the main-text material, the appendix should help the reader gain insight into how the list evolved to its present form over the course of the study.

Third, the number of items under an entry from Table 2.1 is in no way related to the entry's ranking. This independence is easy to see by examining the list. For example, **Fixed-Wing Aircraft** has one of the largest number of entries under it and is ranked BC, whereas **Mine Warfare** has only two entries and is ranked A1.

This appendix also identifies, in the right-hand column of the list, Naval acquisition programs, the lead organization of joint programs, the source of the list entry, etc., whichever is appropriate. For example, item AAA, under **Amphibious Vehicles**, the second **Capability** entry below, is a Naval acquisition program, as indicated by the entry "PRG" at the right. Definitions of all entries in the right-hand column are given in Table A.1. Identifying all Naval acquisition programs as such permits easy identification of possible program impacts associated with various candidate decisions about priorities. The longest list, Table A.2, follows.

Table A.1

Definitions of Lead or Source in Longest List

Abbreviation	Definition
PRG	A Navy or Marine Corps acquisition program
NRL	A Naval Research Laboratory program as given in the *Naval Research Laboratory: Management Brief* (U.S. Department of the Navy, NLCCG, 1992)
Reliance	A Tri-Service Science & Technology Reliance program (under the U.S. Department of Defense, Joint Directors of Laboratories)
USA	A joint program run by the Army
USAF	A joint program run by the Air Force
USMC	A joint program run by the Marine Corps
USN	A joint program run by the Navy
OSD	A joint program run by the Office of the Secretary of Defense
Ad Hoc	Not a program

Table A.2

Longest List of Capabilities and Programs, by RD&T Capability Listed in Table 2.1

RD&T Capability/Program	Lead/Source
Air-Vehicle Signature Control and Management—Stealth features of air platforms.	
F/A-18 E/F Hornet	PRG
JAST, Joint Advanced Strike Technology Program	USAF
JSOW, Joint Stand-Off Weapon	USN
Low-Observables Technology	NRL
Signature Measurement	NRL
Tomahawk, Cruise Missile	PRG
TSSAM, Tri-Service Standoff Attack Missile	USA
Amphibious Vehicles—vehicles designed to operate in the water and on the land. Hovercraft and surface-effects vehicles are included.	
AAA, Advanced Amphibious Assault Vehicle	PRG
Anti-air Weapons—anti-air missiles for all platforms, and point-defense guns.	
AIM-7P Sparrow Block 1, Air Intercept Missile	PRG
AIM-9X, Advanced Sidewinder Missile	USN
AIM/RIM-7P Sparrow PIP	PRG
AMRAAM, Advanced Medium-Range Air-to-Air Missile	USAF
Anti-Air Missiles	Reliance
Avenger, Pedestal-Mounted Stinger	USA
ESSM, Evolved Sea Sparrow Missile	PRG
MHIP, Missile Homing Improvement Program	PRG
Missile Propulsion	Reliance
Phalanx Close-In Weapon System	PRG
RAM, 5-in. Rolling-Airframe Missile	PRG
SM-2 Block IV	PRG
SM-2, Standard SAM Blocks I/II/III	PRG
Stinger RMP, Reprogrammable Microprocessor	USA
Tartar SM-2/NTU Block I/II, New Threat Upgrade	PRG
Terrier SM-2/NTU, New Threat Upgrade	PRG
Anti-ship Weapons—missiles for use (perhaps not exclusively) against ships.	
Anti-Surface Air-Launched Missiles	Reliance
Missile Propulsion	Reliance
Penguin Missile	PRG
Ballistic-Missile Defense—Navy programs contributing to development of ballistic-missile-defense capability.	
Cooperative Engagement Capability	PRG
Missile Propulsion	Reliance
Sea-Based Theater Ballistic Missile Defense	PRG
Ballistic Missiles—programs contributing to Navy ballistic-missile capabilities.	
Missile Propulsion	Reliance
SLBM Retargeting System	PRG
Trident II, Sea-Launched Ballistic Missile	PRG
Biotechnology—programs so designated, plus efforts in antifouling and related areas.	
Chemical/Biological Sensors	NRL

Table A.2—continued

RD&T Capability/Program	Lead/Source

Carrier-Unique Aspects of Fixed-Wing Aircraft—unique aspects of aircraft design and maintenance relating to operation aboard aircraft carriers. This category does not include high-speed aerodynamics or weapons.

AEW, Advanced Early Warning	PRG
Carrier Aircraft Unique	Reliance
EA-6B ADVCAP (RPG, Receiver Processor Group)	PRG
EA-6B Prowler	PRG
F-14 Block 1 Strike Upgrade	PRG
F-14D Tomcat	PRG
F/A-18 C/D Hornet	PRG
F/A-18 E/F Hornet	PRG
JAST, Joint Advanced Strike Technology Program	USAF
T-45TS, Undergraduate Pilot-Training System	PRG

Chemical/Biological Defense

Chemical/Biological Sensors	NRL

Civil Engineering—construction and related technologies; direct combat is covered under **Combat Engineering**

Airfields and Pavements	Reliance
Conventional Facilities	Reliance
Critical Air Base Facilities/Recovery	Reliance
Fire Fighting	Reliance
Ocean and Waterfront Facilities and Operations	Reliance
Survivability and Protective Structures	Reliance
Sustainment Engineering	Reliance

Combatant Ships—hull forms and propulsion-machinery technology.

CVN-68 Nimitz-Class Carrier	PRG
DDG-51, Aegis Destroyer	PRG
LCAC, Landing Craft Air Cushion	PRG
LHD-1, Amphibious Assault Ship	PRG
LX, Amphibious Assault Ship	PRG
MCM-1, Mine Countermeasures Ship	PRG
TAO-187, Fleet Oiler	PRG
Ship Technology	Reliance

Combat Engineering—explosive ordnance disposal, underwater demolition, obstacle breaching, and other combat engineering technologies.

ACS-DEMIS, Advanced Countermeasures System (classified)	USA
Land Mines and Demolition	Reliance

Communications—transmission/reception equipment and terminals and the theory of their operation.

Antijam Communication Links	NRL
Communications Signal Processors	Reliance
Distributed Information Systems	Reliance
Dynamic Spectrum Management	Reliance
JTIDS, Joint Tactical Information Distribution System	USAF
NTCS-A, Naval Tactical Command System Afloat	PRG
NTDS, Naval Tactical Data System Improvements	PRG
Radio Technology	Reliance

Table A.2—continued

RD&T **Capability**/Program	Lead/Source
Radios and Links	Reliance
SINCGARS, Single Channel Gnd & Air Radio—VHF	USA

Computer Science & Technology—processing hardware, data transmission in networks, and software.

AI (Artificial Intelligence)/Neural Networks	Reliance
AN/UYK-43(V) Standard Hardware System	PRG
AN/UYK-44(V) Standard Hardware System	PRG
C2P, Command and Control Processor	PRG
Distributed Processing/High-Performance Computing	Reliance
Enhanced Modular Signal Processor	PRG
Expert Systems	NRL
HCI, Human/Computer Interfaces	Reliance
High-Speed Networking	NRL
Human/Computer Interfaces	NRL
Methods of Specifying and Developing Navy Software	NRL
Networks	Reliance
Parallel Computing Algorithms	NRL
Parallel Processing Algorithms	NRL
Pattern Recognition	NRL
Software and System Engineering	Reliance
Standard Hardware, Environments, Operating Systems	NRL
Trusted Systems & Computer Security	Reliance
Visualization of Scientific Processes	NRL

Conventional-Weapons Effects

Directed-Energy Weapons

Charged-Particle Devices	NRL
Chemical Lasers	NRL
Directed-Energy Effects	NRL
High-Energy Lasers	NRL
High-Power Microwave Sources	NRL
Hydrogen Lasers for GPS	NRL
Laser Fusion	NRL
Laser Propagation	NRL
Pulse Power	NRL

Electronic & Photonic Devices—electronic device theory, operation, and manufacture; sensor hardware and theory (excludes processing).

Aircraft (ASW, Undersea) Electro-Optics	Reliance
Aircraft (Fixed-Wing) Electro-Optics	Reliance
Aircraft (Rotary-Wing) Electro-Optics	Reliance
Battlefield Electro-Optics	Reliance
Electro-Optical Devices	Reliance
Electronic Materials and Processing Science	NRL
Integrated Optics	NRL
Meteorological Effects on Electro-Optics	NRL
Microelectronics	Reliance
Microwave/MM Wave Technology	NRL
Nanoelectronics and Microelectronics	NRL
Radiation-Hardened Electronics	NRL

Table A.2—continued

RD&T **Capability**/Program	Lead/Source
RF Components	Reliance
Shipboard (Low-Elevation) Electro-Optics	Reliance
Specifications and Standards Technology	Reliance
Superconductivity	Reliance
Vacuum Electronics and Microelectronics	NRL
Wide-Area Surveillance (Space-Based IR)	Reliance
Electronic Warfare—offensive and defensive systems, and their management.	
AIEWS, Advanced Integrated EW System	PRG
Aircraft Electronic Warfare	Reliance
ALR-67(V)2 Radar Warning Receiver	PRG
ASPJ, Advanced Self-Protection Jammer	USN
Battle Management Software	NRL
Coherent Countermeasures	PRG
Combat Support Electronic Warfare	Reliance
Communications Jamming	NRL
Decoys (RF & IR)	NRL
EA-6B ADVCAP (RPG)	PRG
EA-6B Prowler	PRG
EW/C3CM (Command, Control, and Communications Countermeasures) System Concepts	NRL
Ground Vehicles Electronic Warfare	Reliance
HARM, High-Speed Anti-Radiation Missile	USN
Maritime Electronic Warfare	Reliance
Repeaters/Jammers, EO/IR	NRL
Signature Measurement	NRL
Environmental Quality—includes quality and science areas.	
Atmospheric Compliance	Reliance
Base Support Operations	Reliance
Cold-Region Science	Reliance
Global Marine Compliance	Reliance
Installation Restoration	Reliance
Lower-Atmosphere Sciences	Reliance
Noise Abatement	Reliance
Ocean Sciences	Reliance
Pollution Prevention	Reliance
Space/Upper-Atmospheric Sciences	Reliance
Terrestrial and Aquatic Assessment	Reliance
Terrestrial Sciences	Reliance
Fixed-Wing Aircraft—Military adaptations of airframes and propulsion, not including carrier- or VSTOL-specific features.	
Aerodynamics	Reliance
AEW, Advanced Early Warning	PRG
CID, Combat ID/Cooperative Aircraft ID	PRG
Combustion Dynamics	NRL
Crew Station	Reliance
EA-6B ADVCAP (RPG)	PRG
EA-6B Prowler	PRG
F-14 Block 1 Strike Upgrade	PRG
F-14D Tomcat	PRG

Table A.2—continued

RD&T **Capability**/Program	Lead/Source
F/A-18 C/D Hornet	PRG
F/A-18 E/F Hornet	PRG
Flight Dynamics/Controls	Reliance
Generic Structures Technology	Reliance
Integrated Avionics System Architecture	Reliance
JAST, Joint Advanced Strike Technology Program	USAF
JPATS, Joint Primary Aircraft Training System	USAF
Land-Based Support Systems	Reliance
Life-Support Systems	Reliance
NAS, National Aerospace System	USAF
P-3 ASUW Improvement Modifications	PRG
P-3 ASUW Program	PRG
P-3 Sustained Readiness Program	PRG
Ramjet Dynamics	NRL
Subsystems	Reliance
Supersonic Hypersonic Flows	NRL
Turbine Engines	Reliance
T-45TS, Undergraduate Pilot-Training System	PRG

Flexible Manufacturing

Fuels and Lubricants
Lubricants and Greases	NRL

Ground Vehicles—emphasis on combat vehicles.
Combat Vehicles	Reliance
Countermine Equipment	Reliance
Material-Handling Equipment	Reliance
MTVR, Medium Tactical Vehicle Replacement	USMC
Power	Reliance
Ramps and Bridging	Reliance

High-Performance Offensive Guns
Conventional Guns	Reliance
Gun-Munitions Safe and Arm	Reliance
NSFS, Naval Surface Fire Support	PRG

Human Factors
Air Quality in Confined Spaces	NRL
Clothing, Textiles, and Food	Reliance
Human/Computer Interfaces	NRL

Information Management—generation, display, and management of military-specific databases; dissemination of command-and-control directives.
Advanced Field Artillery Tactical Data System	USA
CCS MK2 SSN (Combat Control System Improvement)	PRG
Combat Management Information Systems	NRL
Cooperative Engagement Capability	PRG
Data Fusion	Reliance
Decision Aids	Reliance
JTIDS, Joint Tactical Information Distribution System	USAF

Table A.2—continued

RD&T **Capability**/Program	Lead/Source
MIDS-LVT, Multi-functional Information Distribution Terminal	USN
Network Architectures	NRL
NTCS-A, Naval Tactical Command System Afloat	PRG
NTDS, Naval Tactical Data System Improvements	PRG
OSS, Operations Support System	USA
TAOM, Tactical Air Operations Module	USMC
Tactical Decision Aids/Weapon Assessment	NRL

Maintainability, Reliability, Survivability—programs so designated, plus live-fire testing, shock testing, and engineering investigation technology.

CASS, Consolidated Automated Support System	PRG
Chemical/Biological Sensors	NRL
Coatings	NRL
Fire Safety	NRL
Laser Hardening	NRL
Satellite Survivability	NRL

Manpower and Personnel

Force Management and Modeling	Reliance
Human Resources Development	Reliance
Productivity Measurement/Enhancement	Reliance
Selection and Classification	Reliance

Mapping, Geodesy, and Weather

Airborne Geophysical Studies	NRL
JSIPS, Joint Services Imagery Processor System	USAF
Meteorological Nowcasting/Forecasting	NRL
Numerical Weather Prediction	NRL

Marine Mammals

Materials

Advanced Alloy Systems	NRL
Advanced Polymers	NRL
Bio/Molecular Engineering	NRL
Ceramics and Composite Materials	NRL
Corrosion Science	NRL
Energetic Materials Physics	NRL
High-Temperature Materials	NRL
Materials Processing	NRL
Metamorphic Materials	NRL
Rapid-Solidification Technology	NRL
Superconductivity	NRL
Thin Films and Coatings	NRL
Water Additives and Cleaners	NRL
Advanced Materials	Reliance

Mine Warfare—deep, shallow, and surf-zone mine warfare. Does not include underwater demolition teams (see **Combat Engineering**).

MCM-1, Mine Countermeasures Ship	PRG
MHC-51, Coastal Minehunter	PRG

Table A.2—continued

RD&T **Capability**/Program	Lead/Source
Naval Oceanography—emphasis on product requirements to support Naval operations.	
Advanced Electro and Acoustic Mapping	NRL
Aerogeophysics	NRL
Air-Sea Interaction	NRL
Bathymetric Technology	NRL
Benthic Processes	NRL
Bio-Dynamics	NRL
Bio-Optical Models	NRL
Digital MC&G Database Design	NRL
Marine Sedimentary Processes	NRL
Mesoscale Ocean Dynamics	NRL
Ocean Dynamics and Predictions	NRL
Seismic and Acoustic Seafloor Interactions	NRL
Upper-Ocean Dynamics	NRL

Navigation

Navy Medical—Navy medical research areas. Excludes general medical knowledge and research.	
Combat Casualty Care—Blood Research	Reliance
Human Systems Technology	Reliance
Deep-Diving Research	Ad Hoc

Navy Sensors and Signal Processing—sonars, surface search, and low-angle airborne maritime surveillance sensors and processing. Sensors in this category are designed to operate in the water, on the surface, or in the presence of sea-clutter.

Acoustic Transducer and Array Development	NRL
Acoustic Transducer Measurement Technology	NRL
ADS, Advanced Deployable Surveillance	PRG
Aegis Weapon System Modifications	PRG
Aircraft (ASW) Radar	Reliance
Aircraft (ASW, Undersea) Electro-Optics	Reliance
ALFS, Airborne Low-Frequency Sonar	PRG
FDS, Fixed Distributed System	PRG
Low-Frequency Active Acoustics Technology	NRL
MHIP, Missile-Homing Improvement Program	PRG
Point Defense	NRL
Shallow-Water ASW	Ad Hoc
Shallow-Water Environmental Acoustics/Sensors	NRL
Shipboard (Low-Elevation) Electro-Optics	Reliance
SQUID for Magnetic-Field Detection	NRL
Surface Ship ASW Combat System	PRG
SURTASS, Towed Array Sonar (passive)	PRG
SURTASS LFA, SURTASS Low-Frequency-Active	PRG
Undersea Surveillance System Performance	NRL
Underwater Acoustics, Including Propagation	NRL

Non-Maritime Sensors and Signal Processing—overland and high-angle over-water sensors and processing.

Aegis Weapon System Modifications	PRG
AIEWS, Advanced Integrated EW System	PRG
Aircraft (Fixed-Wing) Electro-Optics	Reliance

Table A.2—continued

RD&T **Capability**/Program	Lead/Source
Aircraft (Fixed-Wing) Radar	Reliance
Aircraft (Rotary-Wing) Electro-Optics	Reliance
Aircraft (Rotary-Wing) Radar	Reliance
Battlefield Radar	Reliance
Counter–Low-Observables Technology	NRL
EA-6B ADVCAP (RPG)	PRG
Electromagnetic Sensors—Gamma Ray to RF	NRL
Enhanced Modular Signal Processor	PRG
F/A-18 Radar Upgrade	PRG
Fiber-Optic Sensor Technology	NRL
Imaging Radars	NRL
Low-Observables Technology	NRL
Over-the-Horizon Radar	NRL
Target Classification/Identification	NRL
Ultra-Wide-Band Radar	NRL
Wide-Area Surveillance Radar	Reliance
Nuclear Power/Propulsion	
CVN-68 Nimitz-Class Carrier	PRG
NAS, New Attack Submarine	PRG
SSN-688, Los Angeles–Class Submarine	PRG
Trident Submarine, Ohio Class	PRG
Nuclear-Weapons Effects	
Atmospheric Effects	Reliance
Basic Radiation Hardening	Reliance
Blast/Shock/Thermal Hardening	Reliance
Land Mobile/Fixed Facilities EMP Hardening	Reliance
Missiles/Aircraft EMP Hardening	Reliance
Nuclear-Weapons-Effects Simulation Technology	Reliance
Radiation Hardening Applied Technology	Reliance
Radiation-Hardened Electronics	NRL
Precision Strike—homing weapons and strike-platform enhancements.	
Anti-Armor Weapons	Reliance
Anti-Surface Air-Launched Missiles	Reliance
Anti-Surface Surface-Launched Missiles	Reliance
Bomb Safe and Arm	Reliance
F-14 Block 1 Strike Upgrade	PRG
Guidance and Control	Reliance
Hard-Target Penetrators	Reliance
Javelin, Advanced Anti-Tank Missile	USA
JDAM, Joint Direct Attack Munitions	USAF
JSOW, Joint Stand-Off Weapon	USN
Laser Hellfire	USA
Missile Propulsion	Reliance
MLRS, Multiple-Launch Rocket System	USA
NSFS, Naval Surface Fire Support	PRG
SLAM, Standoff Land Attack Missile	PRG
SLAM-I, Standoff Land Attack Missile, Improved	PRG
Tomahawk, Cruise Missile	PRG
TOW 2, Anti-Tank Missile	USA
TSSAM, Tri-Service Standoff Attack Missile	USA

Table A.2—continued

RD&T **Capability**/Program	Lead/Source
Rotary-Wing Aircraft	
Aerodynamics	Reliance
AH-1, Sea Cobra	PRG
CH-53E, Super Sea Stallion Helicopter	PRG
Crew Station	Reliance
Flight Controls	Reliance
LAMPS MK III, Block II Upgrade	PRG
Medium-Lift Replacement	PRG
Rotorcraft Power Drive Systems	Reliance
SH-60B, LAMPS MK III Helo	PRG
SH-60F, Carrier ASW Helo	PRG
Structure	Reliance
Subsystems	Reliance
Simulation and Modeling—programs that involve advancement of the state of the art in simulation and modeling.	
NTCS-A, Naval Tactical Command System Afloat	PRG
Sea-Based Theater Ballistic Missile Defense	PRG
Space Systems	
Advanced Space Systems	NRL
Artificial Intelligence Applications to Satellites	NRL
Astrometry	Reliance
Electromagnetic Background in Space	NRL
Flight Experiments	Reliance
Guidance, Navigation, and Control	Reliance
Ionospheric and Magnetospheric Modification	NRL
MILSTAR	USAF
Navigation and Time Technology	NRL
NAVSTAR GPS	USAF
NESP, National EHF Satellite Communications Program	PRG
Power	Reliance
Propulsion	Reliance
Remote Sensing, Calibration and Research	NRL
Satellite Analysis	NRL
Satellite Communications	NRL
Satellite Ground Station Design	NRL
Satellite Survivability	NRL
Sea-Launched Booster Technology	NRL
Solar Activity	NRL
Space Recovery Technology	NRL
Space Segment	Reliance
Space Sensing Technology and Applications	NRL
Spacecraft Design, Engineering, Integration	NRL
Spacecraft Materials Technology	NRL
Spacecraft Power Systems	NRL
Structures	Reliance
Survivability	Reliance
Thermal Control	Reliance
UHF Follow-on Satellite Communication System	PRG
Wide-Area Surveillance (Space-Based IR)	Reliance

Table A.2—continued

RD&T **Capability**/Program	Lead/Source
STOVL Aircraft—Short Take-Off Vertical Landing aircraft. Includes jet and tilt-rotor aircraft.	
AV-8B Remanufacture	PRG
JAST, Joint Advanced Strike Technology Program	USAF
Medium-Lift Replacement	PRG
V-22 Osprey, Joint Advanced Vertical Aircraft	PRG
Submarine Communications	
Submarine Communications	Reliance
Submarine Technology—Submarine theory, design, and manufacturing, excluding nuclear power.	
Anechoic Coatings	NRL
NAS, New Attack Submarine	PRG
SSN-21, Seawolf Submarine	PRG
SSN-688, Los Angeles–Class Submarine	PRG
Trident Submarine, Ohio Class	PRG
Unstructured Grids and Unsteady CFD (Computational Fluid Dynamics)	NRL
Support Ships—noncombatant supply, logistics, and prepositioning ships.	
AOE 6	PRG
LSD-41, Cargo Variant	PRG
Ship Technology	Reliance
Strategic Sealift	PRG
Watercraft Technology	Reliance
Training	
Air Crew Training Effectiveness	Reliance
Classroom Instruction	Reliance
Intelligent Computer-Aided Training	Reliance
Land Warfare/Rotary-Wing Training	Reliance
Sea Warfare Training	Reliance
Training Devices and Features	Reliance
Unit Collective Training	Reliance
Undersea Weapons Technology—torpedoes and torpedo defenses.	
MK-48 ADCAP/Modernization	PRG
MK-50 Torpedo	PRG
SSTD, Surface Ship Torpedo Defense	PRG
Underwater Weapons	Reliance
US/UK Surface Ship Torpedo Defense System	PRG
VLA, Vertical Launch ASROC (Anti-surface Rocket)	PRG
Unmanned Air Vehicles (UAV)	
CRUAV, Close-Range UAV	OSD
SRUAV, Short-Range UAV	OSD
Unmanned Ground Vehicles (UGV)	
Tactical Unmanned Ground Vehicle	USA
Unmanned Undersea Vehicles (UUV)	
SOMSS, Submarine Offboard Mine Search System	PRG
Undersea Autonomous Vehicles	NRL

Table A.2—continued

RD&T **Capability**/Program	Lead/Source
Water-Vehicle Signature Control	
DDG-51, Aegis Destroyer	PRG
Low-Observables Technology	NRL
MCM-1 Mine Countermeasures Ship	PRG
MHC-51 Coastal Minehunter	PRG
Signature Measurement	NRL

ABOUT THE AUTHORS

This appendix is provided primarily because we used Delphi methods in the ranking described in the main text. It is appropriate, therefore, to indicate the background and experience of the ten authors. Of the ten authors, eight participated in one or more of the Delphi sessions; the exceptions are Birkler and Bracken. All of the eight but Heppe participated in the final full-day ranking session. Nevertheless, Heppe's views are reflected in the Delphi results.

Six of the report's authors are members of the RAND staff:

Bruno Augenstein, a member of RAND's Defense and Technology Planning Department, has 50 years' experience in the aerospace industry, government, and academia. He has spent about 30 years at RAND, in two periods, the first beginning in 1949. He was an assistant director and special assistant for Intelligence and Reconnaissance in OSD from 1961 to 1965, receiving DoD's Distinguished Public Service Medal in 1965. He chaired a committee designated by the Department of the Navy to review and assess the Navy's medical needs and the Naval Medical Laboratories in 1975. He received an M.S. in aeronautics and physics from Cal Tech in 1945 and is an active participant in a wide range of RD&T matters.

John Birkler is the associate director of RAND's Acquisition and Technology Policy Center. His research spans a multitude of issues on the industrial base, acquisition, system test, and technology, and covers a wide range of systems, such as the B-2, A-12, C-5, C-17, F-117, F/A-18 E/F, Advanced Cruise Missile, Tomahawk, and Los Angeles–class and Seawolf-class submarines. He holds B.S. and M.S degrees in physics and completed the UCLA Executive Program in Management in 1992.

Jim Chiesa, in his capacity as a RAND communications analyst, has participated in the organization and communication of research tasks and results in various areas, but particularly in defense acquisition. He received an M.S. in environmental science from Indiana University in 1977.

Cullen Crain, a recently retired senior member of RAND's Defense and Technology Planning Department, is a member of the National Academy of Engineering and has about 50 years' academic and industrial experience in electronics, telecommunications, nuclear-weapons effects, and related fields. He started at RAND in 1957 and was head of its Engineering and Applied Science Department in the mid-1980s. He received a Ph.D. in electrical engineering from the University of Texas in 1952.

Brian Nichiporuk is a member of RAND's Defense and Technology Planning Department. His recent research has included work on long-term readiness cycles and future technological requirements for U.S. Navy aircraft-carrier, surface-action, and amphibious-warfare groups in the context of possible MRC scenarios. He has also analyzed military organizational cultures and has developed strategies for changing them in periods of political uncertainty. He received a Ph.D. in political science from MIT in 1993.

Ken Saunders, a senior member of RAND's Defense and Technology Planning Department, has over 37 years' experience in the aerospace and undersea-warfare industries, government, and academia. He was an assistant to the Chief Scientist of the Navy's SSBN Security Technology Program in the early 1980s. He received a Ph.D. in engineering (applied mechanics) from the University of California, Berkeley, in 1965.

Four authors are not members of the RAND staff:

Paul Bracken, a RAND consultant, is professor of International Business and a professor of Political Science at Yale University. He was a member of the senior research staff at the Hudson Institute from 1974 to 1983. He received a Ph.D. in operations research from Yale University in 1982. Professor Bracken has 23 years' experience in the defense industry and academia, and writes widely on national security and related matters.

R. Richard Heppe, a RAND consultant, retired as vice president and general manager of Lockheed Aeronautical Systems Company, where he worked from 1947 to 1988. He is a member of the National Academy of Engineering and in 1979 received the National Security Industrial Association's Charles E. Weakley Award for contributions to U.S. ASW posture. He received an Aeronautical Engineering degree from Cal Tech in 1947 and has about 50 years' experience in the aerospace industry.

Richard F. Hoglund is president of Hoglund Technology, Inc., King George, Virginia. He was a division chief for Navy-related programs at ARPA (1972–1975) and a Deputy Assistant Secretary of the Navy, Research and Advanced Systems and, subsequently, Advanced Concepts (1977–1980). He has about 40 years' industrial, government, and academic experience in aerospace and undersea warfare and consults actively in Naval RD&T and in intelligence matters. He received a Ph.D. in engineering (gas dynamics) from Northwestern University in 1960.

Glenn Krumel, a Commander in the U.S. Navy, is currently stationed at the Defense Nuclear Agency's Field Command in Albuquerque, New Mexico. He is an A-6 pilot and participated in Operation Desert Storm while attached to Attack Squadron 185. He was a (Naval) Federal Executive Fellow at RAND during FY 94. He is a graduate of the U.S. Naval Academy and holds an M.S. in engineering management from Catholic University of America.

Abegglen, James C., and George Stalk, Jr., *Kaisha: The Japanese Corporation: How Marketing, Manpower Strategy, Not Management Style, Make the Japanese World Pace-Setters*, New York: Basic Books, 1988.

Ansberry, C., and C. Hymovitz, "Kodak Chief Is Trying for the Fourth Time to Trim Firm's Costs," *Wall Street Journal*, September 19, 1989.

Applegate, Lynda M., and Ramiro Montealegre, "Eastman Kodak Co.: Managing Information Systems Through Strategic Alliances," Cambridge, Mass.: Harvard Business School, Case 9-192-030, August 1993.

Aspin, Les, and General Colin Powell, USA, "Bottom-Up Review," Washington, D.C.: U.S. Department of Defense, briefing charts, September 1, 1993.

Bleil, Rick, "Increasing Competitiveness Through Better Supply Management," *Electronic Business Buyer*, Vol. 19, No. 11, 1993, pp. 72–86.

Bruce, Gregory, and Richard Shermer, "Strategic Partnerships, Alliances Used to Find Ways to Cut Costs," *Oil and Gas Journal*, November 8, 1993, pp. 71–76.

Carroll Publishing Company, *DEFENSE Program Service*, Binders 1 and 2, Washington, D.C., November 1993.

Dalkey, Norman C., *The Delphi Method: An Experimental Study of Group Opinion*, Santa Monica, Calif.: RAND, RM-5888-PR, June 1969.

Defense Science Board, *Report of the Defense Science Board Summer Study Task Force on Defense Manufacturing Enterprise Strategy*, Washington, D.C.: Office of the Under Secretary of Defense for Acquisition (OUSD[A]), September 1993.

——, *Report of the Defense Science Board on Use of Commercial Components in Military Equipment*, Washington, D.C.: OUSD(A), June 1989.

——, *Defense Science Board 1986 Summer Study: Use of Commercial Components in Military Equipment*, Washington, D.C.: OUSD(A), 1987.

Gore, Al, Vice President of the United States, *From Red Tape to Results: Creating a Government That Works Better & Costs Less*, Report of the National Performance Review, Washington, D.C.: U.S. Government Printing Office (GPO), September 7, 1993.

Handy, Charles B., *The Age of Unreason*, Boston, Mass.: Harvard Business School Press, 1991.

Harris, John F., "Defense Cuts: Not in My District," *Washington Post*, June 9, 1994.

Kent, Glenn A., and David E. Thaler, *A New Concept for Streamlining Up-Front Planning*, Santa Monica, Calif.: RAND, MR-271-AF, 1993.

Kettl, Donald F., *Sharing Power: Public Governance and Private Markets*, Washington, D.C.: The Brookings Institution, 1993.

Kleinman, Samuel, and Carla Tighe, eds., *Shrinking the Defense Infrastructure*, Alexandria, Va.: Center for Naval Analyses, Conference Report, 1993.

Kotter, J. P., "What Leaders Really Do," *Harvard Business Review*, May–June 1990, pp. 103–111.

Mills, D. Quinn, *Rebirth of the Corporation*, New York: John Wiley & Sons, 1992.

National Critical Technologies Panel, *Report of the National Critical Technologies Panel*, Washington, D.C., March 1991 (PB91-156869 NTIS).

Olesen, Douglas A., "The Future of Industrial Technology," *Industry Week*, Vol. 242, No. 24, December 20, 1993, pp. 50–53.

Powell, Colin L., Chairman, Joint Chiefs of Staff, *Chairman of the Joint Chiefs of Staff Report on the Roles, Missions, and Functions of the Armed Forces of the United States*, Washington, D.C.: U.S. Department of Defense, CM-1584-93, February 10, 1993.

Prahalad, C. K., and Gary Hamel, "The Core Competence of the Corporation," *Harvard Business Review*, May–June 1990, pp. 79–91.

Ricks, Thomas E., and Michael K. Frisby, "Clinton Nominates Perry for Defense Secretary, but Still Faces Doubts About Strength of His Team," *Wall Street Journal*, January 25, 1994.

Schmitt, Eric, "Cost-Minded Lawmakers Are Challenging a 2-War Doctrine," *New York Times*, March 10, 1994, p. 2.

"Senate Panel Gives Clinton B-2 Funds He Didn't Seek," *Congressional Quarterly*, June 11, 1994.

Slomovic, Anna, *An Analysis of Military and Commercial Microelectronics: Has DoD's R&D Funding Had the Desired Effect?* Santa Monica, Calif.: RAND, N-3318-RGSD, 1991.

Stewart, Thomas A., "Welcome to the Revolution," *Fortune*, December 13, 1993, pp. 66–80.

Towers Perrin, Inc., *Priorities for Competitive Advantage*, New York, 1992.

U.S. Department of Defense, Joint Directors of Laboratories, *Basic Research Panel: Annual Report*, Washington, D.C., June 1993.

———, *Tri-Service Science & Technology Reliance: Annual Report*, Washington, D.C., December 1992.

U.S. Department of the Navy, *RD&A Management Guide*, 12th ed., Washington, D.C.: U.S. GPO, NAVSO P-2457, February 1993.

———, Naval Research Advisory Committee, *Defense Conversion*, Washington, D.C.: U.S. GPO, NRAC 93-1, December 1993.

———, Navy Laboratory/Center Coordinating Group, *Naval Air Warfare Center: Management Brief*, Volume 1 of 5, Washington, D.C., September 30, 1992.

———, *Naval Command, Control and Ocean Surveillance Center: Management Brief*, Volume 3 of 5, Washington, D.C., September 30, 1992.

———, *Naval Research Laboratory: Management Brief*, Volume 5 of 5, Washington, D.C., September 30, 1992.

———, *Naval Surface Warfare Center: Management Brief*, Volume 2 of 5, Washington, D.C., September 30, 1992.

———, *Naval Undersea Warfare Center: Management Brief*, Volume 4 of 5, Washington, D.C., September 30, 1992.

U.S. Navy, "CNO Executive Panel Task Force on Domestic Issues," Washington, D.C.: briefing charts, December 1993.

———, "CNO Executive Panel Task Force on Emerging Technologies: Innovations for Future Naval Warfare," Washington, D.C.: briefing charts, December 1993.

———, "CNO Executive Panel Task Force on Future Naval Forces: 'The Navy After Next'," Washington, D.C.: briefing charts, December 1993.

———, "CNO Executive Panel Task Force on National Security," Washington, D.C.: briefing charts, December 1993.

————, Chief of Naval Operations, *Force 2001: A Program Guide to the U.S. Navy*, Washington, D.C., July 1993.

————, *... From the Sea: Preparing the Naval Service for the 21st Century*, Washington, D.C., September 1992.

————, *Restructuring Naval Forces for New Challenges: The FY95–99 Navy Program Review* and associated briefing charts, Washington, D.C.: PR-95, November 1993.

Venkatesan, Ravi, "Strategic Sourcing: To Make or Not to Make," *Harvard Business Review*, November–December 1992, pp. 98–107.

Watkins, Steven, "Bombs Away! McPeak Eyes the Navy's Air Mission," *Air Force Times*, March 2, 1994.

Williamson, Oliver E., *The Economic Institutions of Capitalism: Firms, Markets, Relational Contracting*, New York: Free Press, 1985.